Contents

GENERAL TESTING

STP 1025

Factors That Affect the Precision of Mechanical Tests

Ralph Papirno and H. Carl Weiss, editors

ASTM
1916 Race Street
Philadelphia, PA 19103

Library of Congress Cataloging-in-Publication Data

Factors that affect the precision of mechanical tests/Ralph Papirno
and H. Carl Weiss, editors.
    Papers of a symposium held in Bal Harbour, FL, on 12–13 Nov.
1987; sponsored by ASTM Committees E-28 on Mechanical Testing, E-24 on Fracture
Testing, and E-09 on Fatigue.
    Includes bibliographies and indexes.
    "ASTM publication code number (PCN) 04-010250-23"—T.p. verso.
    ISBN 0-8031-1251-3
    1. Testing—Congresses.   2. Materials—Testing—Congresses.
I. Papirno, Ralph.   II. American Society for Testing and Materials. Committee E-28 on
Mechanical Testing.   III. ASTM Committee E-24 on Fracture Testing.   IV. ASTM
Committee E-9 on Fatigue.   V. Series: ASTM special technical publication; 1025.
TA410.F33   1989
620.1′1292—dc20   89-34779
                                                                        CIP

NOTE

The Society is not responsible, as a body,
for the statements and opinions
advanced in this publication.

Peer Review Policy

Each paper published in this volume was evaluated by three peer reviewers. The authors
addressed all of the reviewers' comments to the satisfaction of both the technical editor(s)
and the ASTM Committee on Publications.

The quality of the papers in this publication reflects not only the obvious efforts of the
authors and the technical editor(s), but also the work of these peer reviewers. The ASTM
Committee on Publications acknowledges with appreciation their dedication and contribution of time and effort on behalf of ASTM.

Printed in Baltimore, MD
August 1989

Foreword

This publication, *Factors That Affect the Precision of Mechanical Tests,* contains papers presented at the symposium of the same name held in Bal Harbour, Florida on 12–13 November 1987. The symposium was sponsored by ASTM Committees E-28 on Mechanical Testing, E-24 on Fracture Testing, and E-09 on Fatigue. Symposium chairmen were: Roger M. Lamothe, U.S. Army Materials Technology Laboratory; John L. Shannon, Jr., NASA Lewis Research Center; and H. Carl Weiss, Boeing Commercial Aircraft Co. Coeditors of this publication were Ralph Papirno and H. Carl Weiss.

Overview

The practical value of an experiment and the credibility of the results are dependent on the precision and bias present in the process through which the data are acquired. The purpose of this symposium was to serve as a forum for discussing those factors which individually or in total affect the precision of data obtained by mechanical testing methods. Papers were solicited from members of the materials testing community who have experienced problems or concerns in the generation of test data. We are indebted to those presenters who expended the time and effort to share their experiences at Bal Harbor.

This STP has seventeen papers, approximately half the number presented at the symposium. The papers were originally given under different session headings, though much of the material does not easily lend itself to simple categorization. The papers in this STP were screened for contributory value to the science of material testing, and a sincere effort was made to include those providing informative and innovative subject matter.

Hardness Testing

The information in this section deals with several different aspects of hardness testing. A statistical comparison of the results of round-robin Vickers and Knoop hardness testing is offered, showing increased repeatability and reproducibility intervals with increasing specimen hardness and, conversely, improved precision with decreasing hardness and increased test loads. A comparison is given on video image analysis and conventional stage micrometer techniques for microindentation studies, which shows a greater discrepancy with the Knoop over the Vickers' indentation. A discussion of the importance of consistency in test material, test instruments, environmental conditions, and test operator procedures is offered for producing comparable results in hardness test accuracy. Also, a study of gage repeatability and reproducibility is presented, employing the methods of statistical process control (SPC) to interpret equipment, material, and appraiser variables in the results of Rockwell scale test instruments.

Fatigue and Fracture Procedures

This section addresses considerations concerning test methods and instrumentation in fatigue and fracture tests. A study of rapid-loading fracture toughness (J_{Id}) shows for the unloading compliance method, the multiple specimen method, and the electric potential method that J_{Id} is dependent on loading rate for all methods and that the dynamic conditions may be predicted from the static fracture toughness curve. A comparison of crack-following techniques on high-strength aluminum alloy demonstrates that the compliance method and the potential drop method are appropriate for automated crack growth monitoring and that certain errors may be eliminated with calculated correction factors. A paper describing resolution requirements for automated elastic-plastic fracture toughness (J_{Ic}) testing shows that system noise limits the capability of high-resolution analog to digital converters in favor of analog-amplified 12-bit converters. A topic on multiaxial fatigue testing of thin-walled tubular specimens outlines the influencing factors of gage length, specimen geometry, instrumentation, and definition of failure and their affects on the interpretation of test results.

Alignment Problems

This section deals with various material test machine alignment considerations and test specimen gripping configurations. A presentation on potential load frame alignment errors gives requirements for eccentricity, angular deflection, and unit alignment. Good corroborative agreement is found between strain-gaged specimen data and dial indicator, alignment telescope, and electronic clineometer data. A very comprehensive method is presented by which the alignment of flat specimen grips may be checked for errors and improved as necessary. A method is provided to aid in mechanical test setup by quantifying specimen bending loads in pinned clevis fixturing, and finite-element analysis is used to show the importance of uniform load distribution. In addition, concern is expressed for axial or torsional forces present in bending tests, and a description of the details of various three- and four-point loading configurations and the attributes which may affect the precision of pure bending data are included.

General Testing

This section covers a diverse range of subject matter pertinent to the accuracy of mechanical testing. A comparison is made in test machine type versus strain-rate interaction that shows a lower determination of yield strength from servocontrolled test machines. A presentation is given showing the need for recommended standard procedures and reporting consistency in determining a material's resistance to deformation in hot strip rolling. Information is also provided on the use of yield stress pattern phenomena as a quick-look stress indicator. A detailed presentation of test results is given for the determination of wear factors on orthopedic implants by the method of weight loss determination. In addition, an article describing a study involving eight technical journals shows a low incidence of inclusion of precision in measurement data used in the reporting of materials research.

The topics briefly mentioned in this overview are addressed in considerable detail in the following text. While only a few of the unfortunately abundant areas for concern over factors which affect the precision of mechanical tests have been included in this book, it is hoped that this STP will broaden our base of understanding and provide encouragement for more volumes to follow.

I want to thank the authors, the reviewers, the session chairmen, the editors, and the ASTM staff for their combined efforts in bringing this STP to fruition.

H. Carl Weiss
The Boeing Co., Seattle, WA 98124;
symposium cochairman and coeditor

Hardness Testing

George F. Vander Voort[1]

Results of an ASTM E-4 Round-Robin on the Precision and Bias of Measurements of Microindentation Hardness Impressions

REFERENCE: Vander Voort, G. F., **"Results of an ASTM E-4 Round-Robin on the Precision and Bias of Measurements of Microindentation Hardness Impressions,"** *Factors That Affect the Precision of Mechanical Tests, ASTM STP 1025,* R. Papirno and H. C. Weiss, Eds., American Society for Testing and Materials, Philadelphia, 1989, pp. 3–39.

ABSTRACT: An interlaboratory test program was conducted by ASTM Committee E-4 on Metallography, according to ASTM Practice for Conducting an Interlaboratory Test Program to Determine the Precision of Test Methods (E 691), to develop information regarding the precision, bias, repeatability, and reproducibility of measurements of Knoop and Vickers microindentation impressions. Both types of indents were made using loads of 25, 50, 100, 200, 500, and 1000 gf (five of each type at each load) using three ferrous and four nonferrous specimens of varying hardness. The indents were measured by 24 laboratories. Analysis of the test results according to E 691 have been used to prepare a Precision and Bias section for ASTM Test Method for Microhardness of Materials (E 384).

Fourteen laboratories measured the indents in the three ferrous specimens and nine labs had similar Vickers hardness measurements. Of the remaining five laboratories, two were consistently lower while three were consistently higher in measured Vickers hardness. For the Knoop indents in the ferrous specimens, the results were similar except that one lab that got consistently lower Vickers hardnesses had acceptable Knoop hardnesses.

Twelve laboratories measured the indents in the four nonferrous specimens, and the hardness data were in better agreement than for the ferrous specimens due to the much larger indents in the nonferrous specimens. For the Vickers data, one laboratory was consistently lower in hardness while two laboratories were consistently higher in hardness. For the Knoop data, three laboratories were consistently lower in hardness while one laboratory was consistently higher in hardness.

Three laboratories measured both ferrous and nonferrous Vickers and Knoop indents, although one of these laboratories (N) measured only one of the nonferrous specimens. Test results for laboratories N and O were acceptable while those for laboratory M were consistently lower in hardness for all specimens and for both Knoop and Vickers indents. This result suggests a consistent bias either in the calibration or the manner in which the indents were sized.

The repeatability and reproducibility intervals increased with increasing specimen hardness and decreasing test load, that is, with decreasing indent size. The within-laboratory and between-laboratory precision values improved as the specimen hardness decreased and the test load increased, that is, as the indent size increased.

KEY WORDS: microhardness, microindentation hardness, Knoop hardness, Vickers hardness, load, precision, bias, repeatability, reproducibility

The work discussed in this paper can be traced back to the Fall 1972 ASTM E-4 meeting where the decision was made to attempt to develop Knoop to Vickers conversions for

---

[1] Supervisor, Metal Physics Research, Carpenter Technology Corp., Reading, PA 19612–4662.

loads from 25 to 1000 gf. A round-robin interlaboratory program was initiated using steel test blocks from 22.6 to 62.5 HRC that were indented with five Knoop and five Vickers impressions at loads of 25, 50, 100, 200, 500, and 1000 gf. However, only four laboratories completed measurement of the indents. These limited test results showed that there was considerable variability in the measurements of the same indents. This round-robin was abandoned in 1978.

A new round-robin was planned by E-4 in 1980 where the primary objective was the evaluation of indent measurement variability and the secondary goal was to explore the Knoop-Vickers correlation as a function of load. Also, the data obtained in the round-robin could be the basis for developing a Precision and Bias statement for E 384 [1]. However, in the round-robin, each laboratory would only be measuring sets of indents made by one tester. Consequently, this work does not address the added influence of the use of different testers. A round-robin where each laboratory made their own indents was reported by Brown and Ineson [2] which revealed substantial variability. A round-robin of this type should be the subject of future work by E-4.

A round-robin was conducted by ASTM Committee B-4 on Metallic Materials for Thermostats and for Electrical Resistance, Heating, and Contacts between October 1967 and June 1969 to assess the error in measuring Knoop indents in precious metal contact materials; an alloy of Au-22%Ag-3%Ni in wire form was used. Longitudinally mounted wire specimens were indented using a 100-gf load with indents both parallel and perpendicular to the wire axis (15 in each orientation). Twenty-six different people from eleven different laboratories measured the indents. The mean diagonal lengths for the indents perpendicular to the wire axis was about 7% smaller than those parallel to the axis (83.18 versus 89.28 μm), and the standard deviation of the measurements of the perpendicular indents was about 6% greater than that of the parallel indent measurements. For these measurements the extremes in reported hardness were 195.1 to 220.1 HK for the perpendicular measurements (mean of 205.7 HK) and 171.1 to 191.7 HK for the parallel measurements (mean of 178.5 HK).

ASTM Committee B-8 on Metallic and Inorganic Coatings is also conducting a round-robin on electroplated copper and chromium coupons using both Vickers and Knoop indents (made by the round-robin participants) at loads of 25, 50, and 100 gf for the Cu layer and 50, 100, and 200 gf for the Cr layers (two types of Cr-plated specimens were tested). This work was initiated to prepare a Precision and Bias statement for ASTM Test Method for Microhardness of Electroplated Coatings (B 578) [3], and results for the first three samples circulated (a fourth of intermediate hardness is being circulated) have been reported by Horner [4]. Analysis of their data shows a moderate influence of test load on the hardness of the hard chromium platings but no influence of load for the soft copper plating. Data scatter was substantial, with the range of the individual measurements increasing with increasing hardness and decreasing test load. The ranges, as compared to the mean values at each load for each specimen, were slightly greater for the Vickers tests compared to the Knoop tests. The ranges of the data compared to the means averaged 35% for the Knoop tests and 41.9% for the Vickers tests. This degree of variability is rather high but not unusual for microindentation tests, as confirmed by the results of the recent E-4 round-robin discussed in this paper.

Factors Influencing Microindentation Measurement

A number of factors are known to influence the quality of microindentation test results [2,4–11]. These problems relate to either the material being tested, the machine used, or the operator. Separating the individual contributions to measurement variability is quite difficult. Hence, the E-4 round-robin was designed to assess only the errors that would arise in measuring the indents since it was recognized that this was a major source of error.

Due to the nature of the equations used to calculate Vickers and Knoop hardnesses, where the hardness numbers are inversely proportional to the square of the mean diagonal length or the long diagonal, a small error in the diagonal measurement causes a proportionally greater error in the calculated hardness as the indents become smaller. Consequently, users of microindentation tests always try to use the largest possible load and generally try to avoid using loads below 100 gf, if possible.

Because a light microscope is used in the vast majority of cases to measure the indents, most studies of measurement error have concentrated on the influence of the resolution of the optics as the limit to measurement precision. The assumption is made that for each objective the indents will be undersized by a constant amount, irrespective of the size of the indent. However, the magnitude of the measurement error will vary with the numerical aperture (NA) of the objective in an inverse manner, that is, the magnitude of the error decreases with increasing numerical aperture.

Brown and Ineson [2] performed a theoretical analysis of the influence of numerical aperture on measurement errors for both Vickers and Knoop indents. The change in length was calculated based on the assumption that the most accurate measurement was provided by the objective with 1.40 NA. They assumed that the visibility error for the included angle (16°) of the Knoop indent was 7.1 times greater than for the 90° included angle of the Vickers indent. They also performed measurements of two Vickers indents by different operators using two types of measurements: measurement of the length on a projection screen, and measurement with a screw micrometer (with green light and with white light). Their data do not show a clear influence of NA on the Vickers diagonal length. A similar study by Tarasov and Thibault [12] showed a consistent variation in the Knoop diagonal length with NA. However, the change in NA was accompanied by a change in magnification. Brown and Ineson did not report the magnification of their objectives. It is clear that there is need for a more carefully planned experiment on the influence of both NA and magnification on errors in reading both Vickers and Knoop indents. For a constant measurement error (undersized), the measured hardness will increase with decreasing test load, that is, as the indents become smaller.

While these trends are certainly present, there are other factors that can occur and alter this simple analysis. The resolution analysis ignores the larger problem of visibility, which depends on both resolution and image contrast. Visibility is also influenced by the quality of the human eye making the measurement. In general, strong black-white contrast is not a characteristic of microindents, particularly when they are relatively small. Also, for small indents, higher magnification, higher numerical aperture objectives must be used, and image contrast decreases with increasing numerical aperture. Oil immersion objectives provide better image contrast than air objectives of the same numerical aperture and could prove valuable for work with small indents, particularly for specimens with low inherent contrast, for example, ceramics. Thibault and Nyquist [13] reported that diagonal measurements were greater when using oil immersion objectives than dry objectives of the same magnification. Their measurements of indents about 25 μm in length on SiC and Al_2O_3—the latter material has low inherent reflectivity—showed better results for an X54, 1.0 NA oil immersion objective than for an X80, 1.40 NA oil immersion objective due to the better image contrast provided by the X54 objective. The maximum difference in long diagonal length for their measurements was slightly more than 5%, which resulted in about a 12% difference in Knoop hardness. However, such objectives are not available on commercial testers. Other imaging techniques, for example, dark field illumination and differential interference contrast, might prove to be quite useful by providing greater image contrast.

Accurate diagonal measurement is also influenced by the nature of the metal flow around the indents, which can be rather extensive. Measurement of Vickers indents is probably affected more by this problem than measurement of Knoop indents. This distortion makes sizing of the indents more difficult due to poor visibility rather than resolution per se. Consequently, for Vickers indents, when image contrast is impaired, it is likely that there is an equal probability for different operators to either undersize or oversize the indents. This may account for different trends reported for HV versus load (discussed below). On the other hand, the tips of the long diagonal of Knoop indents exhibit poor contrast and are less affected by distortion. Hence, the probability of undersizing Knoop indents is much greater than for oversizing. Consequently, Knoop hardness is always observed to increase with decreasing test load, that is, with decreasing indent size. The results of the present round-robin, as will be shown, suggest that the load-hardness relationship is strongly influenced by image contrast and operator perception of diagonal length rather than material characteristics.

Calibration of the measuring device, usually a Filar micrometer, is another source of error but one that should produce a constant bias by a particular laboratory. Most users employ a standard stage micrometer for calibration. These may be of unknown quality, and the ruling may be a bit coarse for higher magnification calibration. Bergsman [5] reported on use of a stage micrometer made with an optical grating machine of very high accuracy. He found that when a standard ocular screw micrometer was checked against the grating stage micrometer, the calibration factor was not constant.

It is frequently observed, particularly for Vickers indents, that testing produces indents that are highly distorted. That is, Vickers indents are not square but have one or more edges bowed inward or outward. Knoop indents, if the short diagonal is measured as well, sometimes have a long-to-short diagonal ratio quite different than the ideal ratio of 7.114. In such cases, is the measurement of the mean diagonal length, or the long diagonal, proper for the basis of the calculation of the Vickers or Knoop hardness? Measurement of the area of the distorted indent is an alternative approach for determining hardness. At least for the case of bulged indents in cold worked copper, Bergsman [5] has shown that area measurements are not warranted and the diagonal measurement is preferred. This, of course, is only one source for distorted indents and this conclusion may not apply to all cases.

Influence of Load on Hardness

Besides the variability in test results for small diagonal indents, there is the added problem that the hardness is not constant over the range of loads used in microindentation testing. Specifically, for the Vickers test, which produces a geometrically similar indent at all loads, the hardnesses obtained at loads below about 50 gf vary significantly from the relatively constant values obtained at higher loads. A log-log plot of load versus diagonal should exhibit a constant slope, n, of 2 for the full range of loads. If the hardness increases as the load decreases, n will be less than 2.0 while it will be greater than 2.0 if the hardness decreases as the load decreases.

Many studies of the relationship between load and Vickers hardness have been published and can be divided into the following categories with respect to low-load hardnesses:

1. Hardness increases as the load decreases ($n < 2.0$) [5,14–26].
2. Hardness decreases as the load decreases ($n > 2.0$) [7,27–30].

3. Hardness essentially constant with load ($n = 2.0$) [31–33].
4. Hardness increases then decreases with decreasing load [34,35].

In general, trends 1, 2, and 4 become more noticeable as the hardness of the specimen increases, that is, as the indents become smaller.

The Knoop indenter does not produce geometrically similar indents, and the hardness should be expected to vary with load. The degree of variation increases with increasing specimen hardness, that is, with decreasing indent sizes. Because of the visibility problem at the indent tips, the probability of undersizing the indent is far greater than for oversizing. Therefore, there is a very strong agreement that Knoop hardness increases with decreasing load and that the degree of increase raises with increasing hardness [12,13,36–40]. A variation in this trend, where the Knoop hardness increased with decreasing load and then decreased at the lowest loads, was reported by Lysaght [29] and by Blau [33].

Numerous theories have been proposed to explain the load dependence of Vickers hardnesses at low loads. However, none are universally applicable. The results of this round-robin, in which different people measured the same indents, may shed some light on this problem.

E-4 Round-Robin Program

The round-robin discussed in this paper was initiated in November 1980 by M. S. Brooks and R. M. Slepian. Four nonferrous specimens were supplied by M. S. Brooks (J. M. Ney Co., Bloomfield, CT, retired), and three ferrous specimens were supplied by A. R. Fee (Wilson Instrument Division, Binghamton, NY). Table 1 lists the details of these specimens. The polished specimens were indented using five Vickers and five Knoop indents each at loads of 25, 50, 100, 200, 500, and 1000 gf, by A. R. Fee.

Twenty-four people measured the indents. Thirteen people measured all of the indents in the ferrous specimens; one other person measured the indents in only two (F1 and F2) of these specimens. Eleven people measured the indents in the nonferrous specimens; one person measured the indents in only one specimen (NF4). Three people measured both sets of specimens, and, as will be shown, two produced good results while the third did not.

Detailed instructions were given to each person ("laboratory") stating how to perform the measurements and providing a sketch of the specimens showing the location of the indents. A stage micrometer was also circulated with the specimens, and each person was requested to calibrate their microscope against it. Each person was provided with worksheets to record their diagonal measurements for each indent and to list the type of measuring unit used, generally a microhardness tester, the objective magnifications, and the calibration factor for each objective.

TABLE 1—*Test specimens.*

Code No.	Type	Comments
F1	Ferrous	AISI 52100, 63 HRC
F2	Ferrous	AISI 52100, 45 HRC
F3	Ferrous	AISI 52100, 25 HRC
NF1	Nonferrous	44%Pd-38%Ag-1%Ni-16%Cu, annealed and aged
NF2	Nonferrous	44%Pd-38%Ag-1%Ni-16%Cu, annealed
NF3	Nonferrous	75%Au-22%Ag-3%Ni, cold rolled
NF4	Nonferrous	75%Au-22%Ag-3%Ni, annealed

The raw data was summarized by the late A. DeBellis (United Hardness Systems, Inc., retired), who screened out some of the incomplete and outlier data (method not known, probably by inspection). Analysis of the data progressed rather slowly, however. In Spring 1985, the writer volunteered to analyze the ferrous data and, in the Fall of 1985, the nonferrous data. The writer performed this analysis according to ASTM E 691 [41] using the data as summarized by DeBellis. Subsequently, during the preparation of a Precision and Bias statement for E 384 [1], the writer learned of the data not included in the summary tables and was able to obtain this data.

Round-Robin Results

The measurements of each set of five indents for each specimen, at each load, and by each person ("lab") were averaged and are recorded in the upper portion of Tables 2 to 5 by materials type and indenter type. The lower portion of Tables 2–5 contains the hardnesses calculated from the mean diagonal values. For each load, the mean diagonal values and mean hardnesses are listed by the labs (coded alphabetically). The mean diagonal mea-

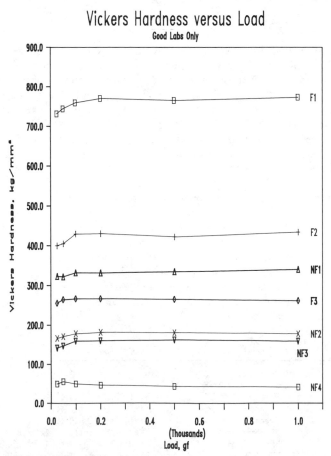

FIG. 1—*Mean Vickers hardness for each test load and specimen.*

surements for each indent type and each material have been averaged in two ways, first for the "good" labs (data summarized by DeBellis) and, second, for all of the labs (including the incomplete data and "outliers"). Note that even for the "good" labs that measured the Vickers and Knoop indents, there are some incomplete data.

Figure 1 shows a plot of the mean Vickers hardness at each load for each specimen for the "good" laboratories. Note that the hardness values are relatively constant at high loads but decrease somewhat at loads below 100 gf. The deviation at low loads increases with increasing specimen hardness, as expected. The one exception is for the softest specimen (NF4), where the hardness rises slightly with decreasing load to 50 gf and then decreases slightly for 25 gf. Figure 2 shows a similar plot for the "good" mean Knoop hardness data. The curves are relatively flat for the low hardness specimens. For the higher hardness specimens, the hardness increases with decreasing load, as expected. This trend is most pronounced for the hardest specimen (F1) with the smallest indents.

Figures 3–5 show the mean Vickers hardness of the "good" labs for each ferrous specimen and the "outlier" and incomplete data for comparison. Figures 6–9 show similar

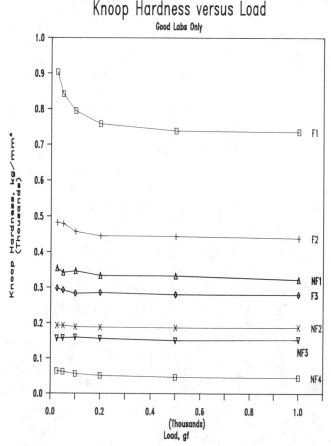

FIG. 2—*Mean Knoop hardness for each test load and specimen.*

TABLE 2—*Vickers hardness and mean diagonal lengths from each laboratory for the ferrous specimens.*

HV - Ferrous Diagonal Measurements, μm

Specimen	Load (gf)	Laboratory ("Good" Labs)									Labs. A to O Mean	Labs. A to O s	Laboratory ("Outlier" Labs.)					Labs. A to J Mean	Labs. A to J s
		A	B	C	D	G	K	L	M	O			E	F	H	N	J		
F1	25	7.76	8.02	7.88	8.28	8.08	8.06	8.14	7.58	7.88	7.96	0.212	6.86	8.48	6.80	9.32	7.30	7.89	0.644
	50	10.98	11.48	10.98	11.36	11.36	11.16	11.36	10.92	10.88	11.16	0.230	10.32	11.58	10.00	12.09	10.50	11.07	0.543
	100	15.62	15.90	15.24	15.60	15.88	15.76	15.90	15.40	15.34	15.63	0.254	14.48	15.95	13.80	16.50	14.80	15.44	0.692
	200	21.78	22.40	21.80	21.78	22.04	21.96	22.42	21.66	21.60	21.94	0.299	20.82	22.30	20.60	22.56	21.20	21.78	0.586
	500	34.82	34.80	34.36	34.84	35.14	35.08	35.16	34.48	34.52	34.80	0.296	33.16	35.20	33.50	36.15	34.20	34.67	0.744
	1000	49.12	48.78	49.00	48.82	49.08	48.80	49.56	48.42	49.00	48.95	0.311	47.08	48.86	46.90	50.40	48.40	48.73	0.885
F2	25	10.62	10.80	10.76	11.12	10.46	11.00	10.70	10.64	10.70	10.76	0.200	10.00	11.19	10.30	11.34	10.06	10.69	0.398
	50	14.84	14.92	15.14	15.44	14.66	15.64	15.22	15.14	15.10	15.12	0.299	14.18	15.49	14.30	16.12	14.62	15.06	0.526
	100	20.64	20.08	21.00	21.36	20.52	21.04	21.16	20.48	20.62	20.77	0.401	19.92	21.37	19.90	21.62	20.42	20.72	0.551
	200	29.36	29.28	29.60	29.80	29.18	29.52	29.48	28.82	28.86	29.32	0.328	28.12	29.58	28.50	30.27	28.98	29.24	0.552
	500	46.48	46.86	47.12	46.84	46.60	47.26	47.04	46.34	47.10	46.85	0.316	45.72	46.93	45.30	47.79	46.11	46.68	0.651
	1000	64.96	65.52	65.60	65.56	65.50	65.40	65.70	64.48	65.00	65.30	0.402	63.92	65.44	63.70	66.12	64.57	65.11	0.704
F3	25	13.62	13.12	13.42	12.90	13.76	14.08	13.60	13.48	13.30	13.48	0.349	12.92	15.29	12.60	13.70		13.52	0.669
	50	18.72	18.22	18.62	18.68	18.62	19.36	19.26	18.78	18.34	18.73	0.374	18.10	19.63	17.60	19.38		18.72	0.579
	100	26.32	26.06	26.32	25.84	26.72	26.76	26.90	26.24	26.30	26.38	0.346	25.90	28.88	25.60	27.09		26.53	0.829
	200	36.98	36.86	37.50	37.44	37.28	37.88	37.68	37.36	37.08	37.34	0.331	36.64	38.79	36.50	38.16		37.40	0.633
	500	59.60	58.88	59.18	58.70	59.36	59.42	59.62	59.00	58.82	59.18	0.343	58.66	60.50	58.00	59.01		59.13	0.602
	1000	84.32	83.30	84.34	83.80	84.12	84.56	85.14	83.70	84.38	84.20	0.546	83.68	85.04	82.70	88.84		84.46	1.479

Vickers Hardness - Ferrous Specimens

Specimen	Load (gf)	Laboratory ("Good" Labs.)									Labs. A to O Mean	Labs. A to O s	Laboratory ("Outlier" Labs.)					Labs. A to J Mean	Labs. A to J s
		A	B	C	D	G	K	L	M	O			E	F	H	M	J		
F1	25	769.9	720.8	746.6	676.2	710.1	713.6	699.7	806.9	746.6	732.26	39.634	985.1	644.7	1002.6	533.7	870.0	759.03	125.761
	50	769.1	703.5	769.1	718.5	718.5	744.5	718.5	777.5	783.3	744.72	30.642	870.6	691.4	927.2	634.3	841.0	761.93	76.861
	100	760.0	733.5	798.4	762.0	735.4	746.6	733.5	781.9	788.0	759.94	24.841	884.4	728.9	973.7	681.1	846.6	782.45	75.313
	200	781.8	739.2	780.4	781.8	763.5	769.1	737.8	790.5	794.9	771.01	20.768	855.6	745.8	874.0	728.7	825.2	783.46	43.324
	500	764.7	765.6	785.4	763.9	750.9	753.4	750.0	779.9	778.1	765.77	13.045	843.2	748.3	826.2	709.5	792.7	772.28	33.530
	1000	768.6	779.3	772.3	778.1	769.8	778.7	755.0	791.0	772.3	773.90	9.798	836.6	776.8	843.1	730.0	791.6	781.66	28.941
F2	25	411.0	397.5	400.4	374.9	423.7	383.1	404.9	409.5	404.9	401.12	14.716	463.6	370.2	437.0	360.5	458.1	407.11	30.721
	50	421.0	416.5	404.5	388.9	431.4	379.1	400.3	404.5	406.6	405.87	15.960	461.1	386.4	453.4	356.8	433.8	410.32	28.607
	100	435.3	459.9	420.5	406.4	440.4	418.9	414.2	442.1	436.1	430.43	16.708	467.3	406.1	468.3	396.7	444.7	432.64	23.025
	200	430.3	432.6	423.3	417.6	435.6	425.6	426.8	446.5	445.3	431.50	9.695	469.0	423.9	456.6	404.8	441.6	434.24	16.549
	500	429.2	422.2	417.6	422.6	427.0	415.1	419.0	431.8	418.0	422.50	5.721	443.6	421.0	451.8	406.0	436.1	425.78	12.011
	1000	439.5	432.0	430.9	431.4	432.2	433.6	429.6	446.0	438.9	434.90	5.405	453.9	433.0	457.0	424.2	444.8	437.64	9.581
F3	25	249.9	269.3	257.4	278.6	244.9	233.9	250.6	255.1	262.1	255.76	13.265	277.7	198.3	292.0	247.0		255.14	23.412
	50	264.6	279.3	267.4	265.7	267.4	247.4	250.0	262.9	275.7	264.48	10.426	283.0	240.6	299.3	246.9		265.40	16.574
	100	267.7	273.1	267.7	277.7	259.7	259.0	256.3	269.3	268.1	266.51	6.979	276.4	222.3	283.0	252.7		264.08	15.341
	200	271.2	273.0	263.7	264.6	266.9	258.5	261.2	265.7	269.7	266.06	4.706	276.3	246.5	278.4	254.7		265.41	8.851
	500	261.0	267.4	264.7	269.1	263.1	262.6	260.8	266.4	268.0	264.81	3.071	269.5	253.3	275.6	266.3		265.22	5.370
	1000	260.8	267.2	260.7	264.1	262.1	258.7	255.8	264.7	260.5	261.62	3.394	264.8	256.4	271.1	235.0		260.15	8.674

TABLE 3—*Knoop hardness and mean diagonal lengths from each laboratory for the ferrous specimens.*

HK - Ferrous Diagonal Measurements, um

Specimen	Load (gf)	A	B	C	D	G	K	L	M	O	Labs. A to O Mean	Labs. A to O s	E	F	H	M	J	Labs. A to J Mean	Labs. A to J s
F1	25	19.34	19.30	19.77	19.64	19.96	20.42	20.04	19.96	20.18	19.85	0.372	18.70	20.13	17.30	21.18	19.30	19.66	0.902
	50	28.40	28.96	28.68	28.82	29.52	29.80	29.34	28.68	29.50	29.08	0.477	27.82	29.39	26.50	29.94	28.50	28.85	0.898
	100	42.02	42.20	42.16	42.22	41.96	42.84	42.72	42.18	42.52	42.31	0.308	41.36	42.71	39.30	43.29	41.70	42.08	0.941
	200	60.74	61.22	60.96	60.46	61.70	61.66	61.80	61.28	61.54	61.26	0.464	60.18	60.91	58.50	62.53	60.70	61.01	0.949
	500	98.62	98.16	99.40	98.70	98.08	98.34	97.80	97.26	96.78	98.13	0.784	96.48	97.61	93.70	103.75	96.90	97.97	2.153
	1000	139.36	139.18	139.44	140.66	138.82	139.74	137.90	138.36	137.82	139.03	0.915	137.08	138.15	134.60	146.43	137.60	138.94	2.596
F2	25	26.86	27.54	27.56	26.70	27.40	27.82	27.04	26.88	26.52	27.15	0.447	26.06	27.34	25.70	27.85	26.26	26.97	0.664
	50	38.56	38.70	38.50	38.06	38.62	39.06	38.54	38.58	38.08	38.52	0.305	37.50	38.41	36.40	39.74	37.59	38.31	0.786
	100	55.78	56.78	56.18	54.90	55.80	56.22	55.46	55.62	55.12	55.76	0.580	54.22	55.72	53.50	56.65	54.74	55.48	0.915
	200	80.12	80.64	80.22	79.32	80.20	81.10	79.52	79.62	78.56	79.92	0.757	78.84	80.07	77.70	80.98	78.77	79.69	0.977
	500	126.36	127.74	126.62	125.80	127.48	127.16	126.48	125.70	125.70	126.56	0.767	125.68	126.11	123.90	131.72	124.61	126.50	1.822
	1000	179.20	180.88	180.04	179.94	181.20	181.12	180.24	179.88	178.00	180.06	1.010	179.04	179.16	177.00	188.91	178.01	180.19	2.797
F3	25	34.24	34.04	34.08	35.02	34.54	35.38	34.84	34.30	34.54	34.56	0.453	32.80	34.73	33.20	34.72	34.35	34.35	0.711
	50	49.48	49.42	49.60	48.94	48.96	50.02	49.52	49.00	49.28	49.36	0.355	47.42	49.08	45.70	49.99	49.03	49.03	0.954
	100	70.96	71.32	70.92	70.54	70.44	71.58	71.42	70.38	70.78	70.93	0.438	68.58	71.63	67.90	71.47	70.61	70.61	1.143
	200	98.92	99.90	99.96	99.80	99.88	101.32	99.58	99.70	99.34	99.82	0.651	95.72	100.14	96.40	105.29	99.69	99.69	2.267
	500	158.38	159.62	159.52	159.66	160.24	162.02	159.20	158.64	158.26	159.50	1.149	157.52	159.60	155.50	166.07	159.56	159.56	2.477
	1000	224.42	225.80	225.28	225.12	224.74	227.58	226.46	225.02	224.56	225.44	1.022	220.00	225.66	221.50	235.27	225.49	225.49	3.537

Columns A–O under Laboratory "Good" Labs.; columns E–J under Laboratory "Outlier" Labs.

Knoop Hardness - Ferrous Specimens

Specimen	Load (gf)	Laboratory ("Good" Labs.)									Labs. A to O Mean	Labs. A to O s	Laboratory ("Outlier" Labs.)					Labs. A to J Mean	Labs. A to J s
		A	B	C	D	G	K	L	N	O			E	F	H	M	J		
F1	25	951.05	954.99	910.13	922.21	892.88	853.11	885.77	892.88	873.52	904.06	34.062	1017.26	877.86	1188.56	792.98	954.99	926.30	92.708
	50	882.08	848.30	864.94	856.56	816.42	801.15	826.46	864.94	817.52	842.04	27.555	919.24	823.65	1013.10	793.67	875.90	857.42	56.709
	100	805.86	799.00	800.52	798.25	808.17	775.31	779.67	799.76	787.02	794.84	11.496	831.79	780.04	921.27	759.28	818.28	804.59	38.342
	200	771.36	759.31	765.80	778.52	747.54	748.51	745.12	757.82	751.43	758.38	11.560	785.78	767.06	831.56	727.83	772.37	765.00	24.441
	500	731.50	738.37	720.06	730.31	739.58	735.67	743.82	752.10	759.58	739.00	11.834	764.31	746.72	810.34	660.95	757.70	742.21	31.873
	1000	732.65	734.55	731.81	719.17	738.36	728.67	748.25	743.28	749.12	736.21	9.670	757.23	745.54	785.39	663.61	751.51	737.80	26.551
F2	25	493.06	469.01	468.33	498.99	473.82	459.62	486.52	492.33	505.79	483.05	15.879	523.80	475.90	538.58	458.63	515.85	490.02	24.502
	50	478.49	475.03	479.98	491.14	477.00	466.32	478.98	477.99	490.63	479.51	7.609	505.92	482.23	536.96	450.49	503.50	485.33	20.402
	100	457.32	441.35	450.83	472.10	456.99	450.19	462.61	459.95	468.33	457.74	9.494	484.01	458.30	497.13	443.38	474.86	462.67	15.498
	200	443.33	437.63	442.22	452.31	442.44	432.68	450.04	448.91	461.11	445.63	8.466	457.84	443.88	471.37	433.96	458.65	448.31	11.070
	500	445.58	436.00	443.75	449.56	437.78	439.99	444.73	450.27	450.27	444.22	5.372	450.41	447.35	463.45	410.05	458.18	444.81	12.422
	1000	443.10	434.90	438.97	439.46	433.37	433.75	438.00	439.75	449.09	438.93	4.955	443.89	443.29	454.18	398.72	449.04	438.54	12.978
F3	25	303.42	307.00	306.28	290.06	296.45	284.18	293.06	302.36	298.17	297.89	7.735	330.65	294.92	322.73	295.09	301.88	301.88	12.842
	50	290.59	291.30	289.19	297.04	296.80	284.35	290.12	296.31	292.96	292.07	4.188	316.39	295.35	326.22	284.69	296.26	296.26	12.026
	100	282.58	279.74	282.90	285.96	286.77	277.71	278.96	287.26	284.02	282.88	3.486	302.54	277.32	308.63	278.57	285.51	285.51	9.570
	200	290.83	285.15	284.81	285.72	285.26	277.21	286.99	286.30	288.37	285.63	3.686	310.60	283.78	306.23	256.70	286.77	286.77	12.845
	500	283.62	279.23	279.59	279.10	277.08	271.02	280.71	282.70	284.05	279.68	3.986	286.73	279.30	294.23	257.97	279.64	279.64	8.466
	1000	282.52	279.08	280.37	280.77	281.72	274.73	277.45	281.02	282.17	279.98	2.522	293.99	279.42	290.02	257.06	280.02	280.02	8.531

TABLE 4—*Vickers hardness and mean diagonal lengths from each laboratory for the nonferrous specimens.*

HV - Nonferrous Diagonal Measurements, µm

Specimen	Load (gf)	Laboratory ("Good" Labs.)									Labs. O to X		"Outlier" Labs.(1)			Labs. O to N	Labs. O to N
		O	P	Q	R	S	T	V	W	X	Mean	s	M	U	N	Mean	s
NF1	25	11.94	11.96	12.16	12.20	12.06	11.44	11.90	11.80	12.34	11.98	0.262	12.61	11.67		12.01	0.322
	50	16.86	17.00	16.98	17.28	16.88	16.12	17.02	17.14	17.34	16.96	0.355	17.81	16.25		16.97	0.472
	100	23.66	23.72	23.82	23.38	23.66	22.78	23.90	23.68	24.00	23.62	0.361	24.36	22.93		23.63	0.455
	200	33.34	33.44	33.46		33.50	32.50	34.00	33.52	33.84	33.45	0.443	34.13	32.64		33.44	0.526
	500	52.64	52.26	52.46	51.26	52.94	51.30	52.84	54.12	53.74	52.62	0.962	53.80	52.02		52.57	0.955
	1000	74.04	74.16	74.04	70.32	74.54	72.54	74.06	75.32	75.44	73.83	1.563	78.78	73.46		74.25	2.056
NF2	25	16.58	16.62	16.62	16.90	16.80	16.32	16.56	17.14	16.84	16.71	0.238	17.81	18.24		16.95	0.581
	50	23.38	23.28	23.14	23.38	23.60	22.60	23.24	23.70	23.40	23.30	0.315	23.60	23.66		23.36	0.311
	100	32.36	32.24	32.56	30.22	32.96	32.00	32.84	32.84	32.58	32.29	0.836	33.31	32.82		32.43	0.818
	200	46.28	45.24	45.26	42.60	46.02	45.10	45.74	45.74	45.98	45.33	1.099	46.45	45.88		45.48	1.047
	500	72.70	72.52	71.62	66.66	73.34	72.42	72.78	71.86	72.78	71.85	2.014	73.60	71.69		72.00	1.878
	1000	102.06	102.04	101.68	95.66	108.28	101.96	103.06	102.86	103.86	102.38	3.235	101.53	101.39		102.22	2.918
NF3	25	17.80	17.84	18.70	17.88	18.00	18.30	17.86	18.46	17.92	18.08	0.323	18.33	18.64		18.16	0.338
	50	25.02	24.72	25.32	26.52	24.90	25.24	24.62	25.06	24.74	25.13	0.573	25.62	24.71		23.04	7.275
	100	34.30	34.02	35.02	32.40	34.22	33.84	34.38	34.76	34.18	34.12	0.740	34.73	34.37		34.20	0.689
	200	48.74	47.78	49.32	46.36	48.70	47.54	48.86	48.40	48.32	48.22	0.886	48.77	48.51		44.28	13.964
	500	75.30	75.84	75.64	73.64	76.22	75.30	75.92	76.80	76.60	75.70	0.930	75.95	77.80		75.91	1.045
	1000	108.08	108.18	107.62	106.66	108.28	107.60	108.32	109.18	108.52	108.05	0.704	108.32	110.67		108.31	1.007
NF4	25	30.64	29.70	30.28	28.88	30.74	29.92	30.96	30.98	31.78	30.43	0.849	31.47	32.46	30.32	30.68	0.965
	50	41.54	40.76	41.10	35.94	41.58	40.86	41.40	42.08	42.28	40.84	1.905	42.31	43.37	40.89	41.18	1.816
	100	61.44	60.14	61.60	58.32	61.54	59.86	61.66	62.06	62.12	60.97	1.269	62.36	63.70	50.63	61.29	1.387
	200	89.16	88.86	89.90	85.44	90.24	88.76	90.46	90.46	90.58	89.32	1.619	91.45	90.28	89.82	89.62	1.526
	500	147.02	145.50	149.04	141.00	147.14	145.28	146.54	149.00	148.94	146.61	2.549	147.37	146.00	145.47	146.53	2.218
	1000	214.24	210.06	212.60	205.20	211.86	209.44	212.28	217.48	214.04	211.91	3.467	211.55	207.91	207.33	211.17	3.393

Vickers Hardness - Nonferrous Specimens

Specimen	Load (gf)	Laboratory ("Good" Labs.)									Labs. O to X Mean	Labs. O to X s	"Outlier" Labs.			Labs. O to N Mean	Labs. O to N s
		O	P	Q	R	S	T	V	W	X			M	U	N (1)		
NF1	25	325.2	324.1	313.5	311.5	318.7	354.2	327.4	333.0	304.4	323.56	14.499	291.6	340.4		322.18	17.232
	50	326.2	320.8	321.6	310.5	325.4	356.8	320.1	315.6	308.4	322.82	14.137	292.3	351.1		322.62	18.250
	100	331.3	329.6	326.8	339.2	331.3	357.4	324.6	330.7	321.9	332.54	10.507	312.5	352.7		332.55	13.003
	200	333.7	331.7	331.3		330.5	351.1	320.8	330.1	323.9	331.62	8.984	318.4	348.1		331.95	10.600
	500	334.6	339.5	336.9	352.9	330.8	352.3	332.1	316.6	321.1	335.19	12.265	320.3	342.6		334.52	12.143
	1000	338.3	337.2	338.3	375.0	333.8	352.4	338.1	326.9	325.8	340.63	15.024	298.8	343.6		337.10	18.515
NF2	25	168.6	167.8	167.8	162.3	164.3	174.1	169.1	157.8	163.5	166.14	4.716	146.2	139.3		161.89	10.472
	50	169.6	171.1	173.2	169.6	166.5	181.5	171.7	165.1	169.3	170.84	4.717	166.5	165.6		169.97	4.646
	100	177.1	178.4	174.9	203.1	170.7	181.1	171.9	171.9	174.7	178.21	9.905	167.1	172.2		176.65	9.579
	200	173.2	181.2	181.1	204.4	175.1	182.3	177.3	177.3	175.4	180.80	9.371	171.9	176.2		179.57	8.868
	500	175.4	176.3	180.8	208.7	172.4	176.8	175.0	179.6	175.0	180.00	11.036	171.2	180.4		179.23	10.227
	1000	178.0	178.1	179.4	202.6	158.2	178.4	174.6	175.3	171.9	177.38	11.488	179.9	180.4		177.89	10.336
NF3	25	146.3	145.7	132.6	145.0	143.1	138.4	145.3	136.0	144.4	141.87	4.949	138.0	133.4		140.75	5.182
	50	148.1	151.7	144.6	131.8	149.5	145.5	153.0	147.6	151.5	147.05	6.364	141.3	151.9		146.96	6.169
	100	157.6	160.2	151.2	176.6	158.4	161.9	156.9	153.5	158.7	159.46	7.227	153.7	157.0		158.71	6.712
	200	156.1	162.5	152.5	172.6	156.4	164.1	155.4	158.3	158.8	159.62	6.021	155.9	157.6		159.10	5.520
	500	163.5	161.2	162.1	171.0	159.6	163.5	160.9	157.2	158.0	161.89	4.058	160.7	153.2		160.99	4.472
	1000	158.7	158.5	160.1	163.0	158.2	160.2	158.0	155.6	157.5	158.86	2.079	158.0	151.4		158.11	2.908
NF4	25	49.4	52.6	50.6	55.6	49.1	51.8	48.4	48.3	45.9	50.17	2.841	46.8	44.0	50.4	49.40	3.116
	50	53.7	55.8	54.9	71.8	53.6	55.5	54.1	52.4	51.9	55.97	6.075	51.8	49.3	55.5	55.02	5.614
	100	49.1	51.3	48.9	54.5	49.0	51.8	48.8	48.1	48.1	49.94	2.149	47.7	45.7	50.4	49.44	2.282
	200	46.7	47.0	45.9	50.8	45.5	47.1	45.3	45.3	45.2	46.53	1.762	44.3	45.5	46.0	46.22	1.646
	500	42.9	43.8	41.7	46.6	42.8	43.9	43.2	41.8	41.8	43.17	1.545	42.7	43.5	43.8	43.21	1.342
	1000	40.4	42.0	41.0	44.0	41.3	42.3	41.2	39.2	40.5	41.32	1.367	41.4	42.9	43.1	41.62	1.338

(1) Incomplete data for Lab. N

TABLE 5—Knoop hardness and mean diagonal lengths from each laboratory for the nonferrous specimens.

HK - Nonferrous Diagonal Measurements. µM

Specimen	Load (gf)	Laboratory ("Good" Labs.)									Labs. O to X Mean	Labs. O to X s	"Outlier" Labs. (1)			Labs. O to M Mean	Labs. O to M s
		O	P	Q	R	S	T	V	W	X			M	U	N		
NF1	25	32.16	31.56	32.16	30.62	31.66	30.40	31.60	32.34	32.74	31.69	0.775	33.38	35.10		32.16	1.300
	50	45.54	45.86	45.54	45.44		43.12	46.98	46.18	46.36	45.63	1.138	46.18	48.12		45.93	1.276
	100	65.24	64.26	64.18	63.70		61.66	62.26	65.58	66.22	64.14	1.581	65.37	65.84		64.43	1.530
	200	92.24	93.72	91.68	90.50		90.98	90.88	93.94	95.06	92.38	1.676	98.32	91.89		92.92	2.410
	500	145.20	146.20	147.20	145.40		145.20	145.20	146.00	148.40	146.10	1.161	153.11	146.61		146.85	2.431
	1000	208.60	209.00	210.00	208.80		208.60	208.80	213.20	213.80	210.10	2.151	213.14	209.11		210.31	2.165
NF2	25	43.52	42.78	42.88	42.74	43.64	41.34	42.32	44.38	42.98	42.95	0.860	43.87	48.63		43.55	1.872
	50	61.24	60.42	60.78	60.16	61.82	59.80	60.44		61.02	60.71	0.644	61.89	63.58		61.12	1.100
	100	87.42	86.66	85.30	86.64	87.32	86.70	86.38	87.04		86.68	0.664	88.05	86.81		86.83	0.726
	200	122.60	122.40	121.20	121.40	124.00	123.40	122.80	126.40	122.20	122.93	1.568	124.34	124.91		123.24	1.566
	500	194.60	194.60	194.20	194.20	196.40	195.00	195.00	196.40	193.80	194.91	0.928	194.19	200.83		195.38	2.000
	1000	272.00	276.00	274.40	272.20	276.80	274.00	277.40	279.00	273.20	275.00	2.435	275.27	285.31		275.96	3.790
NF3	25	47.76	47.98	48.22	46.94	47.78	45.76	47.76	49.02	47.60	47.65	0.897	48.39	50.26		47.95	1.131
	50	67.24	66.96	67.30	67.26	67.54	65.38	67.30	69.00	66.50	67.16	0.951	67.32	66.72		67.14	0.863
	100	93.32	95.10	95.02	93.44	95.16	93.06	92.50	97.88	94.32	94.42	1.622	96.23	90.85		94.26	1.918
	200	134.40	133.40	134.60	133.40	135.40	135.40	135.80	138.40	133.80	134.96	1.565	135.00	130.07		134.52	2.033
	500	216.40	217.50	217.80	216.60	217.80	217.20	217.00	220.00	215.40	217.31	1.269	215.98	207.40		216.29	3.184
	1000	299.80	303.00	306.60	304.40	306.40	304.00	306.20	307.60	304.80	304.76	2.362	301.95	292.60		303.40	4.241
NF4	25	74.46	73.18	74.44	73.90	75.30	74.12	73.72	74.58	77.44	74.58	1.231	75.84	89.51	75.10	75.97	4.406
	50	106.00	104.80	106.40	106.40	107.80	106.40	106.00	108.00	109.60	106.82	1.416	108.99	124.36	108.25	108.58	5.162
	100	157.66	158.60	157.20	158.80	159.40	159.00	158.20	161.00	160.80	158.96	1.288	160.02	173.08	157.79	160.13	4.250
	200	236.58	235.00	233.80	235.40	237.00	234.40	235.60	238.40	241.00	236.35	2.233	233.99	244.51	234.09	236.65	3.249
	500	387.60	392.40	387.80	386.20	392.80	391.20	387.80	393.20	395.60	390.51	3.244	396.70	402.57	392.57	392.20	4.649
	1000	559.80	563.60	562.60	566.20		562.60	562.60	566.00	569.60	564.34	3.191	569.81	558.60	564.20	564.30	3.711

Knoop Hardness - Nonferrous Specimens

Specimen	Load (gf)	O	P	Q	R	S	T	V	W	X	Labs. O to X Mean	Labs. O to X s	"Outlier" Labs. M	"Outlier" Labs. U	(1) M	Labs. O to M Mean	Labs. O to M s
NF1	25	343.94	357.14	343.94	379.41	354.89	384.92	356.24	340.12	331.86	354.72	17.669	319.26	288.74		345.50	25.782
	50	343.05	338.28	343.05	344.56		382.64	322.34	333.61	331.02	342.32	17.930	333.61	307.25		337.94	19.335
	100	334.31	344.58	345.44	350.67		374.25	367.08	330.85	324.49	346.46	17.296	332.98	328.24		343.29	16.690
	200	334.48	324.00	338.58	347.46		343.81	344.56	322.48	314.93	333.79	11.990	294.39	337.03		330.17	16.460
	500	337.45	332.85	328.34	336.52		337.45	337.45	333.76	323.06	333.36	5.237	303.49	330.99		330.14	10.468
	1000	327.00	325.75	322.65	326.37		327.00	326.37	313.04	311.29	322.43	6.506	313.22	325.41		321.81	6.550
NF2	25	187.82	194.37	193.47	194.74	186.79	208.15	198.62	180.61	192.57	193.01	7.789	184.83	150.42		188.40	14.602
	50	189.70	194.89	192.59	196.58	186.16	198.95	194.76		191.07	193.09	4.076	185.74	176.00		190.64	6.689
	100	186.19	189.47	195.56	189.56	186.62	189.29	190.70		187.82	189.40	2.933	183.53	188.81		188.75	3.177
	200	189.33	189.95	193.73	193.09	185.08	186.88	188.72	178.12	190.57	188.39	4.711	184.07	182.39		187.45	4.716
	500	187.87	187.87	188.65	188.65	184.44	187.10	187.10	184.44	189.42	187.28	1.774	188.66	176.40		186.42	3.707
	1000	192.33	186.79	188.98	192.04	185.71	189.53	184.91	182.80	190.64	188.19	3.322	187.78	174.80		186.94	5.005
NF3	25	155.95	154.52	152.99	161.45	155.82	169.88	155.95	148.04	157.00	156.84	6.038	151.92	140.82		154.94	7.299
	50	157.36	158.68	157.08	157.26	155.96	166.44	157.08	149.43	160.88	157.80	4.469	156.98	159.82		157.91	4.055
	100	163.39	157.33	157.60	162.97	157.13	164.30	166.30	148.52	159.94	159.72	5.370	153.66	172.39		160.32	6.513
	200	157.55	159.92	157.08	159.92	155.23	155.23	154.31	148.57	158.96	156.31	3.560	156.15	168.21		157.37	4.802
	500	151.93	150.25	149.98	151.64	149.98	150.81	151.09	146.99	153.34	150.67	1.750	152.52	165.40		152.17	4.689
	1000	158.31	154.98	151.37	153.56	151.56	153.97	151.76	150.38	153.16	153.23	2.399	156.06	166.20		154.67	4.467
NF4	25	64.16	66.42	64.20	65.14	62.74	64.75	65.46	63.78	59.32	64.00	2.046	61.85	44.40	63.07	62.11	5.879
	50	63.32	64.78	62.84	62.84	61.22	62.84	63.32	61.00	59.23	62.38	1.634	59.89	46.00	60.71	60.67	4.893
	100	57.24	56.57	57.58	56.43	56.00	56.28	56.85	54.89	55.03	56.32	0.908	55.57	47.50	57.15	55.59	2.685
	200	50.85	51.53	52.06	51.36	50.66	51.80	51.27	50.07	49.00	50.95	0.951	51.98	47.60	51.93	50.84	1.359
	500	47.36	46.20	47.31	47.70	46.11	46.49	47.31	46.02	45.46	46.66	0.774	45.21	43.90	46.16	46.27	1.084
	1000	45.41	44.80	44.95		44.38		44.95	44.42	43.86	44.68	0.504	43.82	45.60	44.70	44.69	0.587

(1) Incomplete data for Lab. M

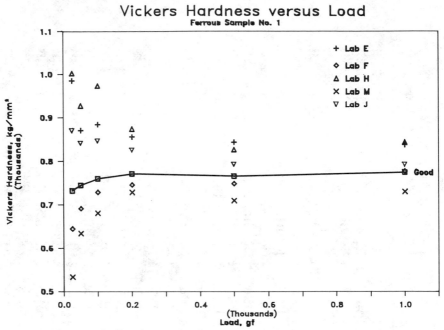

FIG. 3—*Mean Vickers hardness ("good" labs only—squares) versus test load for ferrous sample F1 and "outlier" data (triangle—Lab H, cross—Lab E, inverted triangle—Lab J, diamond—Lab F, X—Lab M).*

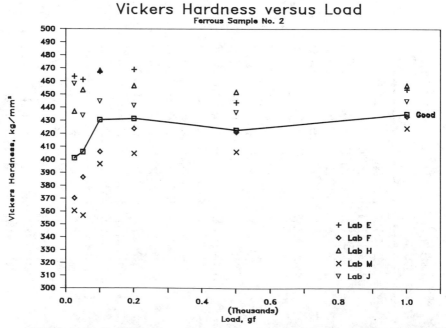

FIG. 4—*Mean Vickers hardness ("good" labs only—squares) versus test load for ferrous sample F2 and "outlier" data (triangle—Lab H, cross—Lab E, inverted triangle—Lab J, diamond—Lab F, X—Lab M).*

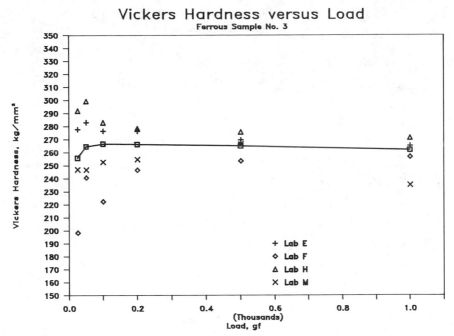

FIG. 5—*Mean Vickers hardness ("good" labs only—squares) versus test load for ferrous sample F3 and "outlier" data (triangle—Lab H, cross—Lab E, diamond—Lab F, X—Lab M).*

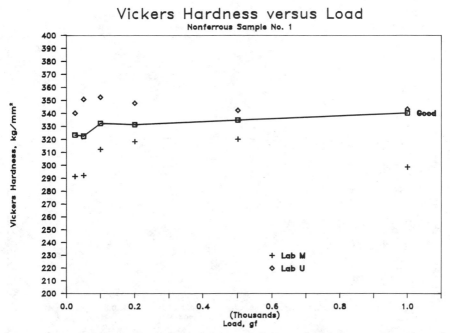

FIG. 6—*Mean Vickers hardness ("good" labs only—squares) versus test load for nonferrous sample NF1 and "outlier" data (diamond—Lab U, cross—Lab M).*

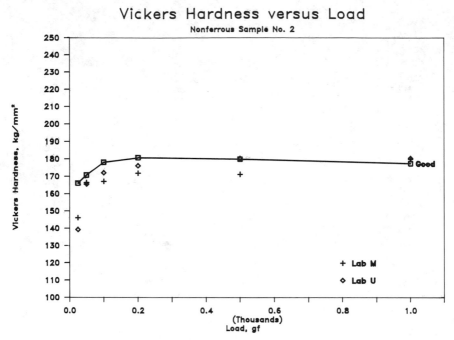

FIG. 7—*Mean Vickers hardness ("good" labs only—squares) versus test load for nonferrous sample NF2 and "outlier" data (diamond—Lab U, cross—Lab M).*

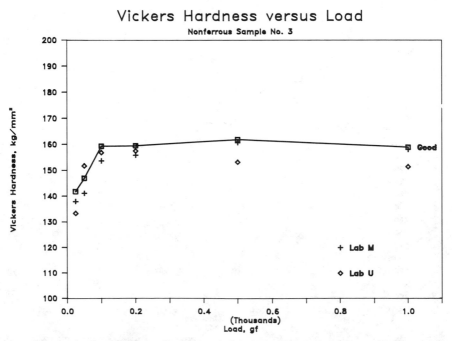

FIG. 8—*Mean Vickers hardness ("good" labs only—squares) versus test load for nonferrous sample NF3 and "outlier" data (diamond—Lab U, cross—Lab M).*

Vickers Hardness versus Load
Nonferrous Sample No. 4

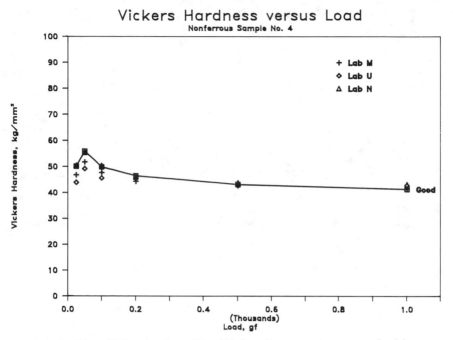

FIG. 9—*Mean Vickers hardness ("good" labs only—squares) versus test load for nonferrous specimen NF4 and "outlier" data (diamond—Lab U, cross—Lab M) and incomplete data for Lab N (triangle).*

Knoop Hardness versus Load
Ferrous Sample No. 1

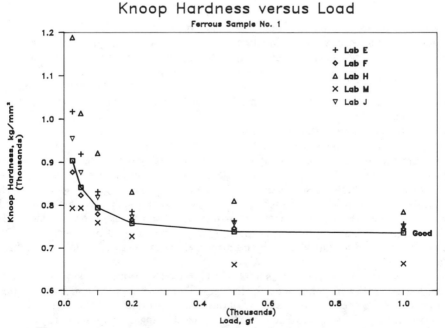

FIG. 10—*Mean Knoop hardness ("good" labs only—squares) versus test load for ferrous specimen F1 and "outlier" data (triangle—Lab H, cross—Lab E, inverted triangle—Lab J, diamond—Lab F, X—Lab M).*

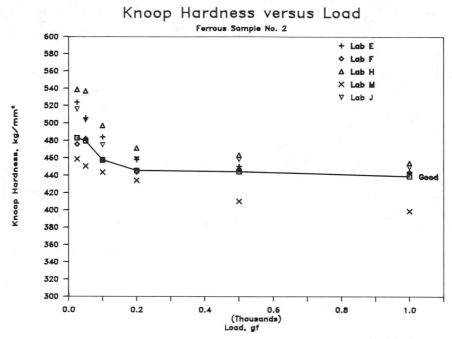

FIG. 11—*Mean Knoop hardness ("good" labs only—squares) versus test load for ferrous specimen F2 and "outlier" data (triangle—Lab H, cross—Lab E, inverted triangle—Lab J, diamond—Lab F, X—Lab M).*

Vickers data for each of the four nonferrous specimens. As might be expected, the degree of data scatter increases with specimen hardness, that is, as the diagonals become smaller. Figures 3–5 are of particular interest because we can see that for the same indents the data for some laboratories (E, H, and J) show an increase in hardness with decreasing load, while others (F and M) exhibit the opposite trend to a greater extent than the mean data. This suggests that the variable trends in HV versus load may be strongly influenced by the measurement process.

Figures 10–12 show corresponding plots of the mean Knoop hardness of the "good" labs and the "outlier" and incomplete data for the ferrous specimens, while Figures 13–16 show similar plots for the nonferrous specimens. Again, the degree of data scatter increases with increasing specimen hardness, that is, with decreasing long diagonal length. Figures 13–16 show some unusual HK versus load trends. Most of lab U's data (diamond symbol) show a decrease in HK at low loads, which, of course, is quite unusual for Knoop data. Lab M's data (crosses) for NF1 are quite erratic, showing a gradual decrease in hardness from 1000 to 200 gf and then an abrupt increase at 100 and 50 gf, followed by a decrease at 25 gf. Otherwise, the data for lab M are in good agreement with the mean values of the "good" data.

To further explore the HV versus load trends, Meyer's hardness type plots of log load versus log diagonal were constructed for the "good" data, for all data, and for the "outlier" data for the seven specimens. The data fall on relatively straight lines but are not shown. The slope of these lines, *n*, was calculated by first calculating a least squares regression line on the log-log data, then calculating the values for 25 and 1000 gf using the regression equation, followed by solving simultaneous equations of the form

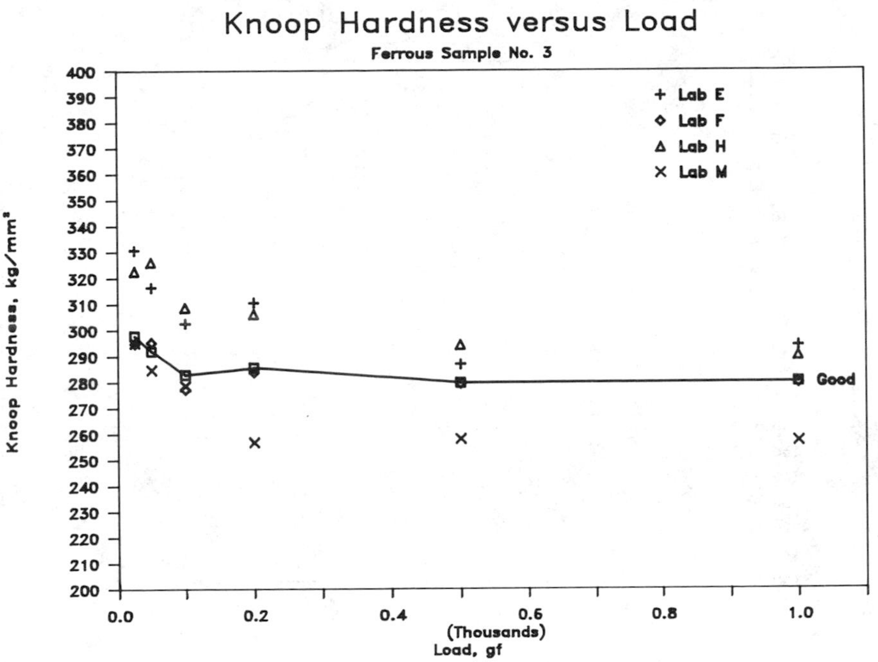

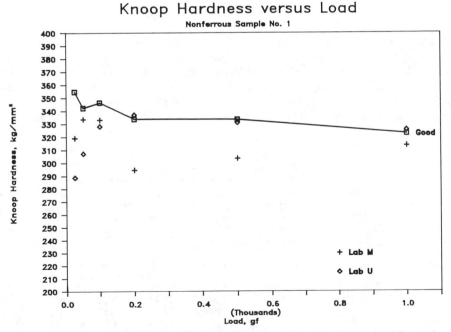

FIG. 12—*Mean Knoop hardness ("good" labs only—squares) versus test load for ferrous specimen F3 and "outlier" data (triangle—Lab H, cross—Lab E, diamond—Lab F, X—Lab M).*

FIG. 13—*Mean Knoop hardness ("good" labs only—squares) versus test load for nonferrous specimen NF1 and "outlier" data (diamond—Lab U, cross—Lab M).*

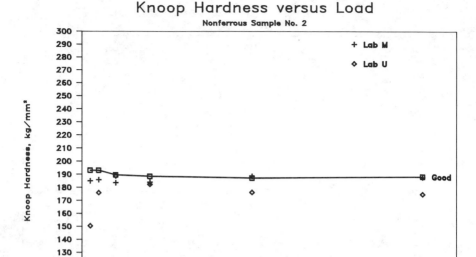

FIG. 14—*Mean Knoop hardness ("good" labs only—squares) versus test load for nonferrous specimen NF2 and "outlier" data (diamond—Lab U, cross—Lab M).*

$$\log \text{load} = \log a + n \log \text{diagonal} \tag{1}$$

The results are given in Table 6. This shows that n is greater than 2 for the six hardest specimens, for the mean diagonals of the "good" labs, and for all labs. This is expected because the plots in Fig. 1 show that the Vickers hardness decreased at loads below 100 gf for the six hardest specimens. For NF4, where Fig. 1 shows the opposite trend, n is less than 2. Note that n for the outlier labs varies considerably with values above and below 2. Note that all of the labs obtained n values below 2 for NF4, the softest specimen.

TABLE 6—*n values (Meyer's exponent).*

Specimen	"Good" Labs	All Labs	Lab E	Lab F	Lab H	Lab M	Lab J
F1	2.029	2.022	1.93	2.094	1.903	2.16	1.949
F2	2.041	2.038	1.98	2.085	2.014	2.097	1.999
F3	2.009	2.010	1.97	2.125	1.953	1.999	NTD

Specimen	"Good" Labs	All Labs	Lab M	Lab U	Lab N		
NF1	2.029	2.026	2.031	1.996	NTD		
NF2	2.037	2.050	2.091	2.126	NTD		
NF3	2.067	2.036	2.086	2.050	NTD		
NF4	1.870	1.882	1.907	1.961	1.885		

NOTE NTD = no test data.

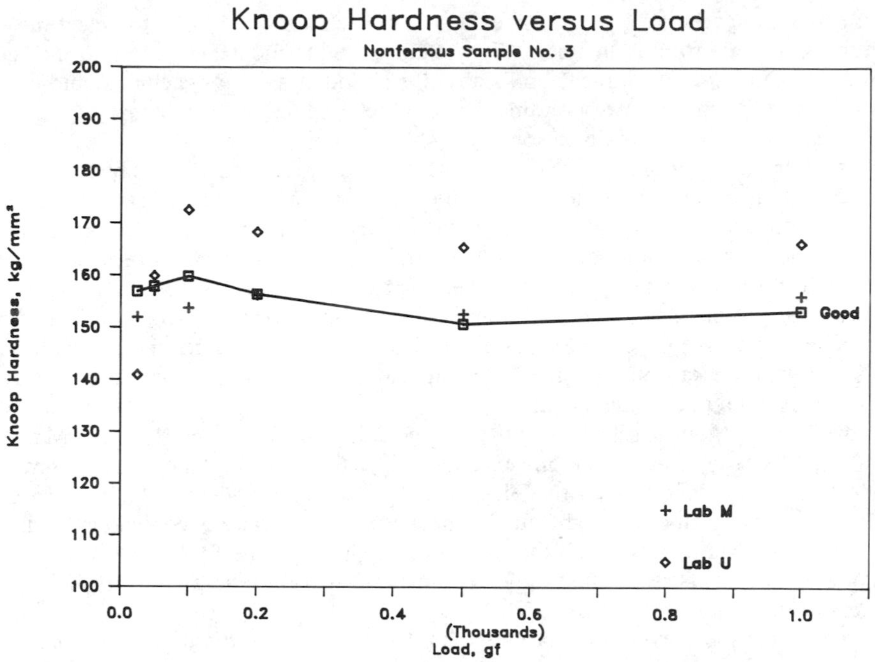

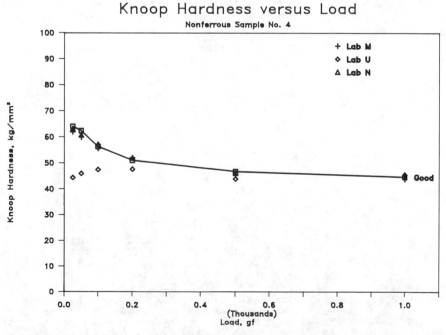

FIG. 15—*Mean Knoop hardness ("good" labs only—squares) versus test load for nonferrous specimen NF3 and "outlier" data (diamond—Lab U, cross—Lab M).*

FIG. 16—*Mean Knoop hardness ("good" labs only—squares) versus test load for nonferrous specimen NF4 and "outlier" data (diamond—Lab U, cross—Lab M) and incomplete data from Lab N (triangle).*

To explore the relative quality of the data from each laboratory and to detect measurement bias, the data from each laboratory was compared to the mean values for the "good" labs. The mean hardness of the "good" labs of each indent type for each load and material was subtracted from the corresponding mean value obtained by each laboratory. The total of the negative and positive deviations from the mean of the "good" labs was determined, along with the number of positive or negative deviations for each lab and by material type (ferrous or nonferrous) and indent type. The sum of the positive and negative deviations was determined and averaged according to the number of materials and loads (a maximum of 18 for the ferrous samples and 24 for the nonferrous samples). Next, the absolute value of these deviations was determined and then the average of the absolute value. The results of these calculations are summarized in Tables 7 and 8 for the ferrous and nonferrous specimens, respectively. Because we do not have an absolute value for the correct hardness measurements, we can only compare our individual data to the mean value of the data that appears to be in best agreement.

Table 7 shows that the selection of the ferrous data from labs E, F, H, J, and M as "outliers" was basically correct. The one exception appears to be the Knoop data from lab F, which are acceptable, although the Vickers data from lab F appear to be biased towards low values. Seventeen of the eighteen HV measurements by lab F were below the mean of the "good" labs while only ten of the eighteen HK values of lab F were below the mean of the "good" labs. This result may indicate that lab F was more familiar with Knoop than Vickers testing.

Table 8 shows less overall deviations than Table 7 because the nonferrous indents are larger and the influence of measurement errors is less. The data in Table 8 reveal that the greatest deviation in the Vickers data was exhibited by lab R, which was not designated as an "outlier." Lab U's Vickers data are really not bad enough to have been rejected but lab U's Knoop data are clearly rejectable. Sixteen of the twenty-four HK nonferrous values by lab U were below the mean. Another reason that lab U's data were rejected was that lab U did not follow the instructions. This person listed hardness values in the raw data table rather than diagonal data as requested. To examine these data, the diagonal values were

TABLE 7—*Deviation of hardness from mean of "good" laboratories—ferrous specimens.*

	Vickers Hardness Data						Knoop Hardness Data				
Lab	No. +	No. −	Mean ΣΔHV	Lab	Mean Σ|ΔHV|	Lab	No. +	No. −	Mean ΣΔHK	Lab	Mean Σ|ΔHK|
M[a]	1	17	−43.57	A	7.7	M[a]	0	18	−34.34	N	5.60
F[a]	1	17	−25.51	C	8.11	K	0	18	−13.75	C	5.78
L	1	17	−12.73	O	9.91	G	6	12	−2.96	C	5.90
K	1	17	−9.54	K	10.08	L	8	10	−2.47	L	6.08
D	9	9	−6.17	G	10.69	F[a]	8	10	−2.21	F[a]	6.63
C	8	10	−2.81	B	11.92	C	9	9	−0.07	B	7.48
B	10	8	−1.18	D	12.74	B	5	13	+0.46	D	8.54
C	8	10	+4.32	L	13.15	O	14	4	+2.86	A	8.91
A	12	6	+5.69	N	13.72	N	15	3	+4.12	O	10.62
O	15	3	+9.11	F[a]	25.84	D	14	4	+4.75	K	13.75
N	14	4	+13.28	M[a]	43.74	A	11	7	+7.07	J**	22.27
J[b]	12	0	+46.02	J[b]	46.02	J**	12	0	+22.27	E[a]	30.04
E[a]	18	0	+57.14	E[a]	57.14	E[a]	18	0	+30.04	M[a]	34.34
H[a]	18	0	+67.62	H[a]	67.62	H[a]	18	0	+62.13	H[a]	62.13

[a] Defined as an "outlier."
[b] Data incomplete, also an "outlier."

TABLE 8—*Deviation of hardness from mean of "good" laboratories—nonferrous specimens.*

	Vickers Hardness Data						Knoop Hardness Data				
Lab	No. +	No. −	Mean $\lvert\Delta HV\rvert$	Lab	Mean $\Sigma\lvert\Delta HV\rvert$	Lab	No. +	No. −	Mean $\Sigma\Delta HK$	Lab	Mean $\Sigma\lvert\Delta HK\rvert$
M[a]	2	22	−9.86	N[b]	0.68	U[a]	8	16	−9.07	N[b]	0.82
X	2	22	−4.91	P	1.63	M[a]	4	20	−7.28	P	1.93
W	2	22	−3.85	O	1.95[a]	W	1	21	−6.25	S	1.97
S	3	21	−2.98	Q	2.72	X	8	16	−4.09	Q	2.13
Q	9	15	−1.83	V	3.19	S	1	18	−1.96	O	2.78
V	6	18	−1.77	S	3.50	O	16	8	−0.24	R	3.64
U[a]	11	13	−0.52	W	4.68	N[b]	3	3	−0.21	V	3.73
O	9	15	−0.50	X	5.50	Q	15	9	−0.08	X	5.38
N[b]	4	2	+0.35	T	7.76	P	15	9	−0.003	W	6.29
P	17	7	+0.52	U[a]	8.01	V	17	7	+1.35	T	7.46
T	20	4	+7.03	M[a]	10.07	R	22	1	+3.59	M[a]	7.88
R	18	5	+8.66	R	12.55	T	17	6	+7.19	U[a]	14.19

[a] Defined as an "outlier."
[b] Incomplete data (only NF4 tested).

calculated from the reported hardnesses. Lab N's data appear to be very good but this person measured only the softest sample, NF4, which was the easiest to measure.

As a final note on the measurement variability, the highest and lowest *individual* measurements of the Vickers and Knoop hardnesses were tabulated for the ferrous specimens (Table 9) and the nonferrous specimens (Table 10). Next to each value, the lab(s) reporting these extreme values has been listed in parentheses. Tables 9 and 10 show that for the

TABLE 9—*Range of individual measurements—ferrous specimens.*

VICKERS HARDNESS (LAB)

Load, gf	F1		F2		F3	
	Min	Max	Min	Max	Min	Max
25	509.4 (M)	1064.3 (H)	338.7 (D)	524.7 (E)	191.2 (F)	311.5 (E)
50	607.9 (M)	965.4 (H)	306.2 (N)	486.9 (E)	215.1 (F)	324.6 (E)
100	653.9 (M)	1032.7 (H)	376.3 (D)	482.7 (E,H)	216.6 (F)	321.9 (D)
200	711.6 (M)	900.0 (H)	394.0 (M)	501.3 (E)	241.7 (F)	281.5 (H)
500	689.5 (M)	861.8 (E)	404.1 (M)	453.8 (E,H)	250.2 (F)	283.4 (H)
1000	716.0 (M)	843.1 (E,H)	413.3 (M)	501.6 (H)	233.9 (M)	275.8 (H)

KNOOP HARDNESS (LAB)

Load, gf	F1		F2		F3	
	Min	Max	Min	Max	Min	Max
25	742.4 (M)	1260.4 (H)	441.0 (C,M)	555.7 (H)	276.0 (D)	338.9 (H)
50	764.8 (M)	1036.4 (H)	439.6 (M)	549.0 (H)	266.8 (M)	352.9 (E)
100	742.4 (M)	950.1 (H)	423.0 (M)	508.5 (H)	272.8 (M)	314.2 (H)
200	711.6 (M)	846.0 (H)	429.5 (M,K)	477.5 (H)	248.8 (M)	322.8 (H)
500	633.5 (M)	819.1 (H)	401.8 (M)	475.7 (H)	283.6 (M)	303.1 (H)
1000	653.2 (M)	790.1 (H)	391.4 (M)	460.5 (J)	249.0 (M)	299.4 (E)

TABLE 10—*Range of individual measurements—nonferrous specimens.*

VICKERS HARDNESS (LAB)

Load, gf	NF1 Min	NF1 Max	NF2 Min	NF2 Max	NF3 Min	NF3 Max	NF4 Min	NF4 Max
25	270.1 (M)	390.2 (T)	129.8 (M)	214.5 (R)	117.1 (P)	172.4 (P)	34.6 (M)	61.3 (R)
50	241.4 (R)	391.0 (T)	142.6 (R)	200.6 (T)	122.6 (R)	163.7 (T)	47.5 (X)	83.1 (R)
100	285.2 (R)	437.0 (R)	162.3 (M)	213.1 (R)	140.0 (W)	200.7 (R)	45.7 (W,U)	59.1 (R)
200	309.8 (M)	357.7 (T)	167.9 (M)	208.3 (R)	108.4 (W)	181.5 (R)	43.5 (V)	54.5 (R)
500	308.8 (W)	384.6 (R)	164.4 (M)	207.8 (R)	152.8 (W)	185.5 (R)	40.6 (Q)	47.1 (R)
1000	291.2 (M)	403.4 (R)	155.5 (S)	204.6 (M)	151.4 (U)	167.9 (R)	38.0 (W)	46.7 (R)

KNOOP HARDNESS (LAB)

Load, gf	NF1 Min	NF1 Max	NF2 Min	NF2 Max	NF3 Min	NF3 Max	NF4 Min	NF4 Max
25	288.8 (U)	414.4 (T)	150.4 (U)	226.8 (V)	140.8 (U)	179.6 (T)	44.4 (U)	72.0 (P)
50	302.5 (V,W,X)	399.5 (T)	176.0 (U)	206.5 (T)	134.6 (W)	183.3 (T)	46.0 (U)	71.1 (P)
100	316.0 (X)	393.9 (V)	178.0 (M)	203.1 (Q)	143.4 (W)	174.9 (R)	47.5 (U)	63.2 (Q,T)
200	290.9 (X)	350.8 (T,V)	171.0 (W)	197.6 (Q, R)	145.2 (V)	168.2 (U)	44.5 (X)	55.3 (N)
500	320.5 (Q, X)	347.9 (T)	176.4 (U)	195.0 (M,Q,X)	141.8 (W)	165.4 (U)	41.9 (M)	50.3 (V)
1000	305.0 (W)	335.3 (O,P)	174.8 (U)	202.6 (O,R)	149.0 (W)	166.2 (U)	41.4 (W,X)	47.7 (O)

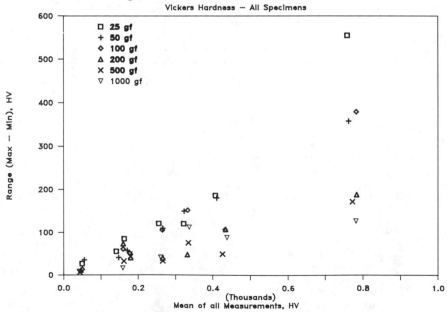

FIG. 17—*Plot of the range of the individual Vickers hardness measurements (all labs) versus the mean Vickers hardness (all labs) of each specimen as a function of test load (square—25 gf, cross—50 gf, diamond—100 gf, triangle—200 gf, X—500 gf, inverted triangle—1000 gf).*

ferrous HV data, labs H and E accounted for nearly all of the high data while labs M and F accounted for most of the low data. For the nonferrous HV data, lab R dominated the individual high data, with lab T next, while labs M, W, and R exhibited most of the individual low data. Note that for the nonferrous HV data, lab R had seventeen of the individual high values and four of the individual low values, which strongly suggests that lab R is out of control. Figure 17 shows a plot of the individual data range for each specimen and test load versus the mean hardness of each specimen (all labs) for the Vickers hardness data.

Tables 9 and 10 show that for the ferrous HK data lab H accounted for most of the individual high data while lab M accounted for most of the individual low data. The distribution of individual high and low nonferrous HK data was more even, with labs T, V, and Q exhibiting the most individual high data and labs U, W, and X exhibiting most of the individual low data. Figure 18 shows a plot of the individual data ranges for each specimen and test load versus the mean hardness of the specimens (all labs) for the Knoop hardness data.

E 691 Analysis

The statistical analysis approach defined in ASTM E 691 [41] was used to define the four following parameters:

1. Repeatability Interval, $I(r)_j$

$$I(r)_j = 2.83(S_r)_j \qquad (2)$$

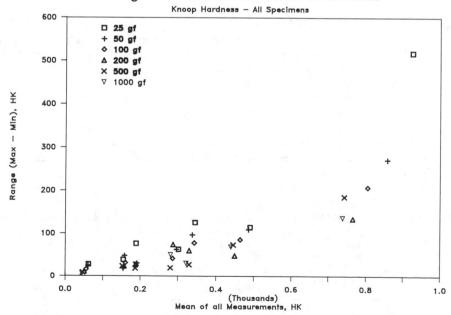

FIG. 18—*Plot of the range of the individual Knoop hardness measurements (all labs) versus the mean Knoop hardness (all labs) of each specimen as a function of test load (square—25 gf, cross—50 gf, diamond—100 gf, triangle—200 gf, X—500 gf, inverted triangle—1000 gf).*

2. Reproducibility Interval, $I(R)_j$

$$I(R)_j = 2.83(S_R)_j \tag{3}$$

3. Precision—within laboratory

$$\pm 2[V_r(\%)]_j \tag{4}$$

4. Precision—between laboratories

$$\pm 2[V_L(\%)]_j \tag{5}$$

The repeatability interval, $I(r)_j$, is the maximum permissible difference due to test error between two test results in the laboratory on the same material. The reproducibility interval, $I(R)_j$, is a comparison of test results obtained in different laboratories on the same material. The within-laboratory precision is for a single operator, same machine, and same day while the between-laboratory precision is for different operators, different machines, and different test dates.

To determine these values, we must first determine $(S_r)_j$, $(S_R)_j$, $[V_r(\%)]_j$, and $[V_L(\%)]_j$. These quantities are defined as:

5. $(S_r)_j$ = The estimated standard deviation within laboratories for material j.
6. $(S_R)_j$ = The reproducibility precision for material j, the square root of the sum $(S_r)_j^2$ + $(S_L)_j^2$, where $(S_L)_j$ is the between-laboratory variance.
7. $[V_r(\%)]_j$ = The estimated coefficient of variation within laboratories for material j.
8. $[V_L(\%)]_j$ = The estimated coefficient of variation between laboratories for material j.

After the values of parameters 5 to 8 were determined, it was noted that the values varied with indent size. $(S_r)_j$ and $(S_R)_j$ varied linearly with the diagonal while $[V_r(\%)]_j$ and $[V_L(\%)]_j$ exhibited a log-log linear relationship with the diagonal. Hence, simple linear regression analysis was used to evaluate these trends before determining parameters 1 to 4. Tables 11 and 12 list these regression equations; $\overline{X}_j$ is the diagonal length (μm) for the j material. Note that a few of the correlation coefficients are rather low, indicating no trend with respect to the diagonal length.

The regression equations were used to compute the repeatability interval, $I(r)_j$, the reproducibility interval, $I(R)_j$, and the within-laboratory and between-laboratory precisions for both Vickers and Knoop data. These data were used to plot the reproducibility and repeat-

TABLE 11—*Regression equations—ferrous specimens.*

VICKERS HARDNESS	
$(S_r)_j = 0.231 + 0.00284\overline{X}_j$	$r = \quad 0.535$
$(S_R)_j = 0.31 + 0.004\overline{X}_j$	$r = \quad 0.747$
$\mathrm{Log}\,[V_r(\%)]_j = 1.109 - 0.735\,\mathrm{log}\overline{X}_j$	$r = -0.861$
$\mathrm{Log}\,[V_L(\%)]_j = 1.04 - 0.715\,\mathrm{log}\,\overline{X}_j$	$r = -0.927$
KNOOP HARDNESS	
$(S_r)_j = 0.216 + 0.006\overline{X}_j$	$r = \quad 0.823$
$(S_R)_j = 0.333 + 0.007\overline{X}_j$	$r = \quad 0.899$
$\mathrm{Log}\,[V_r(\%)]_j = 0.833 - 0.486\,\mathrm{log}\,\overline{X}_j$	$r = -0.846$
$\mathrm{Log}\,[V_L(\%)]_j = 0.852 - 0.55\,\mathrm{log}\,\overline{X}_j$	$r = -0.794$

TABLE 12—*Regression equations—nonferrous specimens.*

VICKERS HARDNESS

$(S_r)_j = 0.373 + 0.008\ \overline{X}_j$ $\qquad\qquad r = 0.862$
$(S_R)_j = 0.0357 + 0.0156\ \overline{X}_j$ $\qquad\qquad r = 0.891$
$\text{Log}\,[V_r(\%)]_j = 1.1 - 0.49\ \log\overline{X}_j$ $\qquad\qquad r = -0.583$
$\text{Log}\,[V_L(\%)]_j = 0.428 - 0.135\ \log\overline{X}_j$ $\qquad\qquad r = -0.20$

KNOOP HARDNESS

$(S_r)_j = 0.057 + 0.0177\ \overline{X}_j$ $\qquad\qquad r = 0.82$
$(S_R)_j = 0.378 + 0.0177\ \overline{X}_j$ $\qquad\qquad r = 0.862$
$\text{Log}\,[V_r(\%)]_j = 0.714 - 0.232\ \log\overline{X}_j$ $\qquad\qquad r = -0.32$
$\text{Log}\,[V_L(\%)]_j = 1.561 - 0.773\ \log\overline{X}_j$ $\qquad\qquad r = -0.807$

ability intervals for the ferrous and nonferrous specimens for both Vickers and Knoop hardnesses as a function of test load, as shown in Figs. 19–22. As might be expected, the reproducibility interval (between laboratories) is greater than the repeatability interval (within laboratory), and these values increase with hardness, that is, with decreasing diagonal length. The data also show that the reproducibility and repeatability intervals are lower for the Knoop hardness test than for the Vickers hardness test. This is probably due to the simple fact that for the same sample and test load, the Knoop indent is larger than the Vickers indent. Also, while errors in measuring the Vickers impression can be either plus or minus, errors in measuring Knoop indents are generally always in the negative direction, that is, shorter than actual size.

The regression equations in Tables 11 and 12 for the coefficients of variation within and between laboratories (parameters 7 and 8, above) were used to calculate the within-labo-

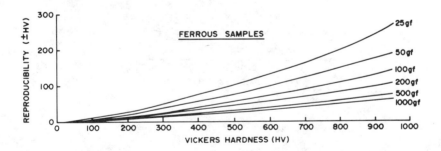

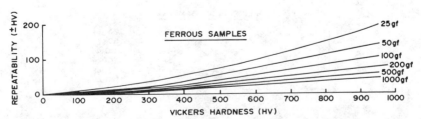

FIG. 19—*Repeatability and reproducibility intervals for the Vickers hardness ($\pm\Delta HV$) for the ferrous specimens as a function of specimen hardness and test load.*

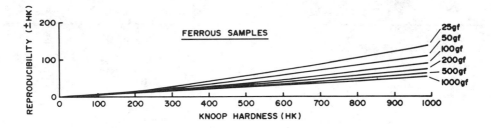

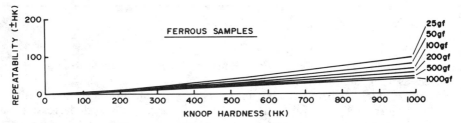

FIG. 20—*Repeatability and reproducibility intervals for the Knoop hardness (±ΔHK) for the ferrous specimens as a function of specimen hardness and test load.*

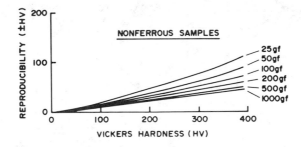

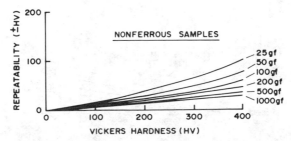

FIG. 21—*Repeatability and reproducibility intervals for the Vickers hardness (±ΔHV) for the nonferrous specimens as a function of specimen hardness and test load.*

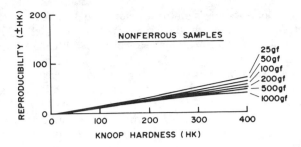

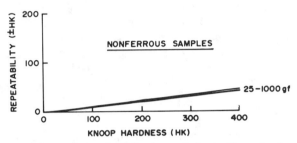

FIG. 22—*Repeatability and reproducibility intervals for the Knoop hardness (± ΔHK) for the nonferrous specimens as a function of specimen hardness and test load.*

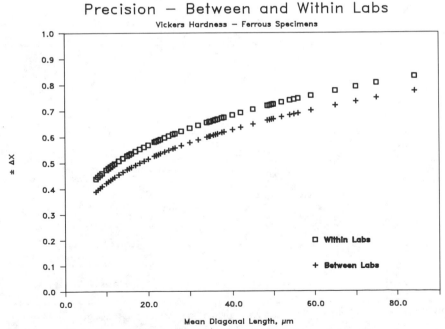

FIG. 23—*The precision in measuring the mean Vickers diagonals (± ΔX) as a function of the mean diagonal length (squares—within lab, crosses—between labs).*

ratory and between-laboratory precision (parameter 3, Eq 4 and parameter 4, Eq 5). However, only the regression equations for the ferrous specimens, which exhibited decent correlation coefficients, were further analyzed. However, even here, as will be shown, the results were conflicting.

The within-lab and between-lab precisions for the ferrous specimens were analyzed in two ways. First, the precision in measuring the diagonal length for the range of diagonals was determined for both the Vickers and Knoop data. Figures 23 and 24 plot the within-lab and between-lab precision values for the ferrous Vickers and Knoop data, respectively. The plots show that in both cases the between-lab precision is lower than the within-lab precision, although the results in both cases are relatively similar. These results suggest that there is probably little difference in the between-lab and within-lab precisions.

The second approach to analyze the precision data was to show how the precision in measuring the Vickers and Knoop diagonals influences the hardness as a function of test load. Figures 25–28 show plots of the within-lab and between-lab precision in hardness as a function of test load for the Vickers and Knoop data, respectively. As expected, the precision in terms of hardness decreases with decreasing test load. Note that the precision in measuring the Knoop diagonals is poorer than for measuring the mean Vickers diagonal; but, because the Knoop diagonal is longer at the same load and for the same specimen, the precision in Knoop hardness is better than the precision in Vickers hardness.

Conclusions

The ASTM E-4 round-robin has confirmed that the reading of the indent size is the major factor influencing the variability of test results. The following conclusions can be drawn:

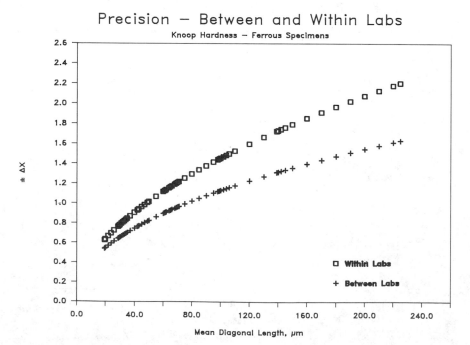

FIG. 24—*The precision in measuring the long Knoop diagonal ($\pm \Delta X$) as a function of the mean diagonal length (squares—within lab, crosses—between labs).*

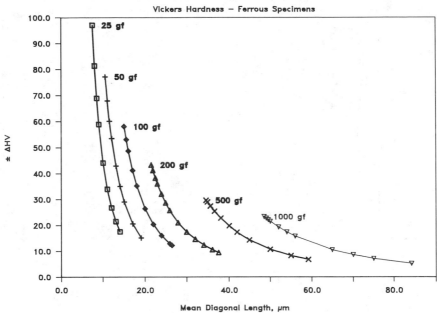

FIG. 25—*The within laboratory precision in terms of Vickers hardness (±ΔHV) as a function of the mean diagonal length and test load (squares—25 gf, crosses—50 gf, diamonds—100 gf, triangles—200 gf, X—500 gf, inverted triangles—1000 gf).*

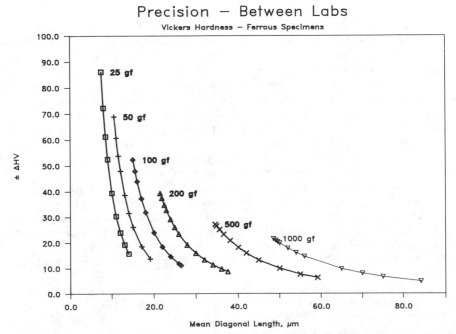

FIG. 26—*The between-laboratory precision in terms of Vickers hardness (±ΔHV) as a function of the mean diagonal length and test load (squares—25 gf, crosses—50 gf, diamonds—100 gf, triangles—200 gf, X—500 gf, inverted triangles—1000 gf).*

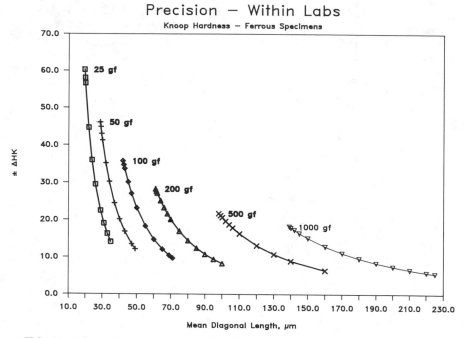

FIG. 27—*The within-laboratory precision in terms of Knoop hardnesses (±ΔHK) as a function of the long diagonal length and test load (squares—25 gf, crosses—50 gf, diamonds—100 gf, triangles—200 gf, X—500 gf, inverted triangles—1000 gf).*

1. The measured Vickers and Knoop hardnesses are relatively constant for loads of 100 gf and greater. For loads below 100 gf, the variation in hardness increases with increasing specimen hardness, that is, with decreasing diagonal length. These trends have been well documented in the literature.

2. Several laboratories produced measurements substantially different from that of most laboratories. The degree of variation of the "outlier" data increased with increasing hardness, that is, with decreasing diagonal length. More "outliers" were exhibited by the ferrous data than the nonferrous data, although different laboratories measured each set.

3. Except for the softest nonferrous specimen, the mean Vickers hardness decreased at loads below 100 gf producing *n* values above 2. The softest nonferrous specimen exhibited a slight increase in hardness and then a slight decrease as the load decreased below 100 gf and the *n* value was below 2. For the outlier data, both increasing and decreasing Vickers hardnesses below 100 gf were observed. These results suggest that the measurement per se of Vickers indents may be a major factor in the low load-hardness variation problem.

4. For the Knoop hardness data, the hardness increased with decreasing test load with the magnitude increasing with increasing hardness. The "outlier" laboratories, however, exhibited either increasing or decreasing Knoop hardnesses with decreasing load, depending on the laboratory.

5. Several of the laboratories exhibited a strong consistent negative or positive bias in measuring the diagonals. In general, the choice of the "outlier" laboratories was correct, except for laboratory R's Vickers data, which were not so classified, and laboratory F's Knoop data, which were acceptable.

6. Examination of the individual test measurements for extreme values (high or low)

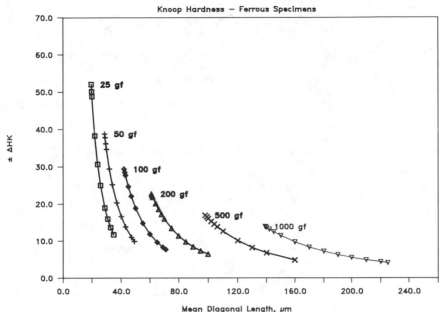

FIG. 28—*The between-laboratory precision in terms of Knoop hardnesses (±ΔHK) as a function of the long diagonal length and test load (squares—25 gf, crosses—50 gf, diamonds—100 gf, triangles—200 gf, X—500 gf, inverted triangles—1000 gf).*

revealed that most were due to the "outlier" laboratories. The range of the individual data was quite high, particularly for the higher hardness specimens and lighter loads.

7. ASTM E 691 was used to calculate repeatability and reproducibility intervals for the ferrous and nonferrous Vickers and Knoop data. Regression analysis was employed to develop these values as a function of hardness and test load. The results show that the reproducibility and repeatability intervals are larger for Vickers data than Knoop data for a given specimen and test load. These values increase with increasing hardness and decreasing load, that is, with decreasing indent size.

8. Within-laboratory and between-laboratory precision estimates were also made. Regression analysis showed a strong correlation between indent size and precision for the ferrous data but inconsistent correlations for the nonferrous data. For the ferrous data, the precision, in terms of the diagonal length, was poorer for the Knoop ferrous data than for the Vickers ferrous data. However, in terms of hardness, the precision of the ferrous Knoop data was better than for the ferrous Vickers data. The between laboratory precision estimates were slightly better than the within laboratory precision estimates, although both were quite similar. This suggests that there is no difference in precision between or within laboratories for microindentation measurements.

References

[1] ASTM E 384-73, "Standard Test Method for Microhardness of Materials," ASTM, Philadelphia.
[2] Brown, A. R. G. and Ineson, E., "Experimental Survey of Low-Load Hardness Testing Instruments," *Journal of the Iron and Steel Institute*, Vol. 169, December 1951, pp. 376–388.

[3] ASTM B 578-80, "Test Method for Microhardness of Electroplated Coatings," ASTM, Philadelphia.

[4] Horner, J. D., "Microhardness Testing of Plated Coatings; Recent Round Robin Experiences," *Testing of Metallic and Inorganic Coatings, ASTM STP 947,* ASTM, Philadelphia, 1987, pp. 96–110.

[5] Bergsman, E. B., "Problems of Hardness Testing of Thin Sheet Metals," *Sheet Metal Industry,* Vol. 31, January 1954, pp. 5–14, 18.

[6] Miodownik, A. P., "The Significance of Microhardness Testing," *Bulletin of the Institute of Metals,* Vol. 2, June 1955, pp. 258–262.

[7] Mott, B. W., *Micro-Indentation Hardness Testing,* Butterworths Scientific Publications, London, 1956.

[8] Bückle, H. "Progress in Micro-Indentation Hardness Testing," *Metallurgical Reviews,* Vol. 4, No. 13, 1959, pp. 49–100.

[9] Petty, E. R., "Hardness Testing," *Techniques of Metals Research,* Vol. V, Part 2, Interscience, New York, 1971, pp. 157–221.

[10] Samuels, L. E., "Microindentation in Metals," *Microindentation Techniques in Materials Science and Engineering, ASTM STP 889,* ASTM, Philadelphia, 1986, pp. 5–25.

[11] Blau, P. J., "Methods and Applications of Microindentation Hardness Testing," *Applied Metallography,* Van Nostrand Reinhold Co., New York, 1986, pp. 123–138.

[12] Tarasov, L. P. and Thibault, N. W., "Determination of Knoop Hardness Numbers Independent of Load," *Transactions of ASM,* Vol. 38, 1947, pp. 331–353.

[13] Thibault, N. W. and Nyquist, H. L., "The Measured Knoop Hardness of Hard Substances and Factors Affecting Its Determination," *Transactions of ASM,* Vol. 38, 1947, pp. 271–330.

[14] Berhardt, E. O., "The Microhardness of Solids in the Boundary Region of the Kick Similarity Principle," *Zeitschrift für Metallkunde,* Vol. 33, 1941, pp. 135–144.

[15] Tate, D. R., "A Comparison of Microhardness Indentation Tests," *Transactions of the American Society for Metals,* Vol. 35, 1945, pp. 374–389.

[16] Onitsch, E. M., "Micro-Hardness Testing," *Mikroscopie,* Vol. 2, 1947, pp. 131–151.

[17] Bergsman, E. B., "Some Recent Observations in Micro-Hardness Testing," *Bulletin American Society for Testing and Materials,* No. 176, September 1951, pp. 37–43.

[18] Bückle, H., "Investigation into the Load Dependency of Vickers Micro-Hardness," *Zeitschrift für Metallkunde,* Vol. 45, November and December 1954, pp. 623–632, 694–701.

[19] Campbell, R. F. "Dependence of Diamond Pyramid Hardness Number on Experimental Variables," *Materials Research & Standards,* Vol. 7, October 1967, pp. 443–449.

[20] Braunovic, M., et al., "Grain-Boundary Hardening in Iron and Iron Alloys," *Metal Science Journal,* Vol. 2, 1968, pp. 67–73.

[21] Braunovic, M. and Haworth, C. W., "The Use of Microhardness Testing to Measure the Thickness of Work-Hardened Surface Layers," *Practical Metallography,* Vol. 7, April 1970, pp. 183–187.

[22] Henderson, E. P., et al., "Microhardness of a Lunar Iron Particle and High-Purity Iron Samples," *The Metallographic Review,* Vol. 1, No. 2, December 1972, pp. 11–14.

[23] Upit, G. P. and Varchenya, S. A., "The Size Effect in The Hardness of Single Crystals," *The Science of Hardness Testing and Its Research Applications,* American Society for Metals, Metals Park, OH, 1973, pp. 135–146.

[24] Bückle, H., "Use of the Hardness Test to Determine Other Material Properties," *The Science of Hardness Testing and Its Research Applications,* American Society for Metals, Metals Park, OH, 1973, pp. 453–494.

[25] Czernuszka, J. T. and Page, T. F., "A Problem in Assessing the Wear Behavior of Ceramics: Load, Temperature and Environment Sensitivity of Indentation Hardness," *Proceedings of the British Ceramic Society,* 1984, Vol. 34, *Ceramic Surfaces and Surface Treatment,* pp. 145–156.

[26] Tirupataiah, Y. and Sundararajan, G., "Evaluation of Microhardness Correction Procedures," *Wear,* Vol. 110, No. 2, 15 July 1986, pp. 183–202.

[27] Campbell, R. F., et al., "A New Design of Microhardness Tester and Some Factors Affecting The Diamond Pyramid Hardness Number at Light Loads," *Transactions of the American Society for Metals,* Vol. 40, 1948, pp. 954–982.

[28] Rostoker, W., "The Effect of Applied Load in Micro-Indentation Tests," *Journal of the Institute of Metals,* Vol. 77, 1950, pp. 175–184, 621–629.

[29] Lysaght, V. E., "The How and Why of Microhardness Testing," *Metal Progress,* Vol. 78, August 1960, pp. 93–97, 122, 124.

[30] Vander Voort, G. F., *Metallography: Principles and Practice,* McGraw-Hill Book Co., New York, 1984, pp. 379–382.

[*31*] Taylor, E. W., "Micro-Hardness Testing of Metals," *Journal of the Institute of Metals,* Vol. 74, 1948, pp. 493–500.
[*32*] Roberts, W., "The Micro-Indentation Hardness of Glazes," *Transactions of the British Ceramic Society,* Vol. 64, 1965, pp. 33–59.
[*33*] Blau, P. J., "A Comparison of Four Microindentation Hardness Test Methods Using Copper, 52100 Steel, and an Amorphous Pd-Cu-Si Alloy," *Metallography,* Vol. 16, 1983, pp. 1–18.
[*34*] Vanchiva, G. V., et al., "Relationship Between Microhardness and Load," *Physics of Metals and Metallography,* Vol. 17, No. 2, 1964, pp. 69–71.
[*35*] Jindal, P. C. and Gurland, J., "An Evaluation of The Indentation Hardness of Spheroidized Steels," *The Science of Hardness Testing and Its Research Applications,* American Society for Metals, Metals Park, OH, 1973, pp. 99–108.
[*36*] Knoop, F., et al., "A Sensitive Pyramidal-Diamond Tool For Indentation Measurements," *Journal of Research National Bureau of Standards,* Vol. 23, July 1939, pp. 39–61.
[*37*] Batchelder, G. M., "The Nonlinear Dispariety in Converting Knoop to Rockwell C Hardness," *Materials Research & Standards,* Vol. 9, November 1969, pp. 27–30.
[*38*] Young, C. T. and Rhee, S. K., "Evaluation of Correction Methods for Determining Load-Independent Knoop Microhardness," *Journal of Testing and Evaluation,* Vol. 6, May 1978, pp. 221–230.
[*39*] Blau, P. J., "Use of a Two-Diagonal Measurement Method for Reducing Scatter in Knoop Microhardness Testing," *Scripta Met.,* Vol. 14, July 1980, pp. 719–724.
[*40*] Hays, C. and Kendall, E. G., "An Analysis of Knoop Microhardness," *Metallography,* Vol. 6, 1973, pp. 275–282.
[*41*] ASTM E 691-79, "Practice for Conducting An Interlaboratory Test Program to Determine the Precision of Test Methods," ASTM, Philadelphia.

Andrew R. Fee[1]

An Alternative Method for Measuring Microindentations

REFERENCE: Fee, A. R., **"An Alternative Method for Measuring Microindentations,"** *Factors That Affect the Precision of Mechanical Tests, ASTM STP 1025,* R. Papirno and H. C. Weiss, Eds., American Society for Testing and Materials, Philadelphia, 1989, pp. 40–45.

ABSTRACT: An image analysis method was investigated to determine if successful measurements of Knoop and Vickers indentations could be realized. Hopefully, this semiautomated method would eliminate operators' fatigue and thus improve the repeatability and reproducibility associated with the conventional manual optical measurement method.

The specimens included a transverse of Vickers indentations of approximately 50 μm in diagonal length. These indents were measured with both methods, and the results compare the conventional optical method with the image analysis findings.

Another technique of calibrating the image analyzer with a known size indentation rather than with a stage micrometer shows a closer agreement to conventional measurement results.

A second study of Knoop and Vickers indentations of varying sizes, from 25 to 98 μm, shows the different results between the two types of indentations and the subsequent discrepancies experienced.

KEY WORDS: image analysis, Knoop indentation, Knoop hardness values, Vickers indentation, Vickers hardness values, image processor, optical microscope, video image, threshold, grey level, pixel array

This study was conducted to explore an alternative semiautomated measuring method for determining Knoop and Vickers hardness values.

This measuring method utilizes image analysis techniques commercially introduced in 1963. Since that time, image analyzers have progressed considerably using special improved scanners that exhibit high resolution, low signal to noise ratios, and accurate control of the scan lines.

An automated measuring technique for microhardness indentations is desirable to improve on the accuracy and reproducibility of conventional optical methods. Knoop and Vickers indentations, defined as feature-specific objects, offer such an opportunity. Objects with specific features are those that exhibit shading differences from the fields background and are easily detected at a particular "threshold" or "grey level."

This grey level determination enables a measuring technique that utilizes quantitative imaging technology and is the method discussed in this study.

---

[1] Technical consultant, Wilson Instrument Division, Page-Wilson Corp., Bridgeport, CT 06602.

Procedure

This study reports measurements on several samples consisting of various size Knoop and Vickers indentations. These indentations were first measured on a conventional microscope using a reflected green filtered light source. In most measurements, the highest magnification of approximately 500 was used.

The image analyzer system used consisted of a solid state camera, video monitor, computer, image processor, and an optical microscope with magnifications of 100, 200, and 500 power. The video image is sent to the image processor from the camera along with picture synchronization information. The image processor passes the image through to the monitor for viewing along with superimposed enhancements that are relevant to analyzing the image. This video image is stored in the processor at a particular "threshold" or "grey level" and detects and subsequently measures the video image.

Results and Discussion

The first series of measurements were done on a tooth profile of carburized 9310 to a depth of 899 μm (0.035 in.). These tests were made with a Vickers indenter at a load of 1000 grams-force, starting 63.5 μm (0.0025 in.) from the edge and transversing inward to the core.

The indentations were measured at 500 magnification with the optical microscope and were plotted (Fig. 1) as the accepted true measurements. Measurements with the image analyzer unit were then made and compared to the former results (Table 1). In both instances, the same optical equipment was used to eliminate one possible variable.

The microscope was calibrated using a Bausch and Lomb precision stage micrometer with a reflected light source so that a calibration could be obtained. The indentations were measured in the prescribed manner, and the results given are the average of both diagonals.

The image analyzer was set up using the same optical microscope with a camera attached to the vertical tube of the microscope. The camera is a solid state Javelin Model 207G with a pixel array of 384 horizontal and 485 vertical. The monitor was a Javelin 9-in., black

TABLE 1—*Vickers indentation—carburized 9310—1000 g load—calibrate with stage micrometers.*

Test No.	X_{cm}	H_{IA}	$X_{cm} - X_{IA}$, μm	HV_{cm}	HV_{IA}	$HV_{IA} - HV_{cm}$, HV	μm, %	HV, %
1	50.2	48.7	1.5	736	782	46	2.9	6.3
2	50.4	49.1	1.3	730	769	39	2.6	5.3
3	50.8	49.1	1.7	719	769	50	3.3	7.0
4	50.7	49.3	1.4	721	763	42	2.8	5.8
5	50.9	49.2	1.7	716	766	50	3.3	7.0
6	51.3	49.8	1.5	705	748	43	2.9	6.1
7	52.1	50.7	1.4	683	721	38	2.7	5.6
8	52.1	50.3	1.8	683	733	50	3.5	7.3
9	52.7	51.0	1.7	668	713	45	3.2	6.7
10	53.6	52.1	1.5	646	683	37	2.8	5.7
			Avg 1.55			44.0	3.0	6.28

NOTE: X = measured indentations in micrometres; H = type; IA = image analysis; HV = hardness Vickers; X_{cm} = measured diagonals with conventional microscope; X_{IA} = measured diagonals with image analysis.

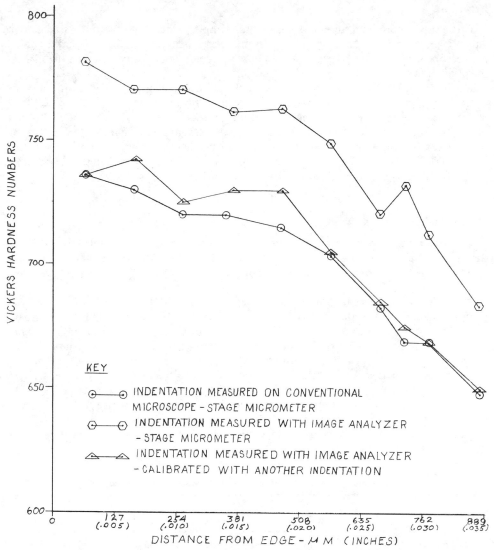

FIG. 1—*Vickers transverse-carburized 9310 1000-g load (data from Tables 1 and 2).*

and white Model BWM-9 with a resolution of 800 lines. An image processor, Model 3000 "Image Technology Disc," and an IBM computer made up the image analyzer system.

A comparison of the results indicates that, although there is a 3% difference in micrometers, the trend in hardness levels is fairly compatible. The average discrepancy of 1.6-μm bias condition is caused when the grey level thresholding is unable to detect the indentations' extreme end points.

It was found that the resolution of the optical microscope seemed slightly better than that obtained with the scanner camera.

TABLE 2—*Vickers indentations—carburized 9310—1000-g load—calibrate with known indentation length.*

Test No.	X_{cm}	H_{IA}	$X_{cm} - X_{IA}$, μm	HV_{cm}	HV_{IA}	$HV_{IA} - HV_{cm}$, HV	μm, %	HV, %
1	50.2	50.2	0.0	736	736	0	0.0	0.0
2	50.4	50.0	0.4	730	742	12	0.8	1.6
3	50.8	50.6	0.2	719	724	5	0.4	0.7
4	50.7	50.4	0.3	721	730	9	0.6	1.2
5	50.9	50.4	0.5	716	730	14	1.0	2.0
6	51.3	51.1	0.2	705	710	5	0.4	0.7
7	52.1	52.0	0.1	683	686	3	0.2	0.4
8	52.1	52.4	0.3	683	675	8	0.6	1.2
9	52.7	52.6	0.1	668	670	2	0.2	0.3
10	53.6	53.5	0.1	646	648	2	0.2	0.3
			Avg 0.22			6.0	0.44	0.84

Another technique used was to calibrate the image analyzer field using one of the optically measured indents as the accepted true measurement (Table 2). This technique compensates for the lesser resolution of the image analyzer system and, as would be expected, has better correlation. This comparison shows a 0.44% difference in micrometers between the conventional optical measurement technique and the image analysis method.

The second series of measurements was done on Knoop and Vickers indentations of comparable length to determine if error with one type was more prominent than with

TABLE 3—*Vickers indentation—steel 52100.*

Test No.	X_{cm}	X_{IA}	$X_{cm} - X_{IA}$, μm	HV_{cm}	HV_{IA}	$HV_{IA} - HV_{cm}$, HV	μm, %	HV, %
			300-G LOAD					
1S	27.0	25.9	1.1	763	829	66	4.1	8.7
2S	26.9	26.0	0.9	769	823	54	3.3	7.0
3S	26.9	25.7	1.2	769	842	73	4.5	9.5
4S	26.8	25.9	0.9	775	829	54	3.4	7.0
5S	26.7	25.7	1.0	780	842	62	3.7	7.9
			Avg 1.02			61.8	3.8	8.02
			1000-G LOAD					
6S	49.0	48.2	0.8	772	798	26	1.6	3.4
7S	48.7	48.1	0.6	782	802	20	1.2	2.6
8S	49.0	48.4	0.6	772	792	20	1.2	2.6
9S	48.9	48.2	0.7	776	798	22	1.4	2.8
10S	49.1	48.1	1.0	769	802	23	2.0	3.0
			Avg 0.74			22.2	1.48	2.88
			3000-G LOAD					
11S	83.9	83.0	0.9	790	808	18	1.1	2.30
12S	83.9	84.1	0.2	790	787	3	0.2	0.03
13S	83.9	84.0	0.1	790	788	2	0.1	0.03
14S	83.9	83.6	0.3	790	796	6	0.4	0.80
15S	84.2	83.8	0.4	785	792	7	0.5	0.90
			Avg 0.38			7.20	0.46	0.07

another (Figs. 1 & 2). The Vickers indents are more readily and accurately measured than are the Knoop indentations. The Vickers errors proportionally decreased with the increase in size, ranging from approximately 26 to 84 μm, whereas the Knoop errors remained somewhat constant, 3 to 4 μm for the same range of length (Tables 3 & 4).

Conclusion

This study indicates that the resolution of the image analysis system with the Vickers indentation results in a measurement of from 0.4 to 1.6 μm shorter than when measured

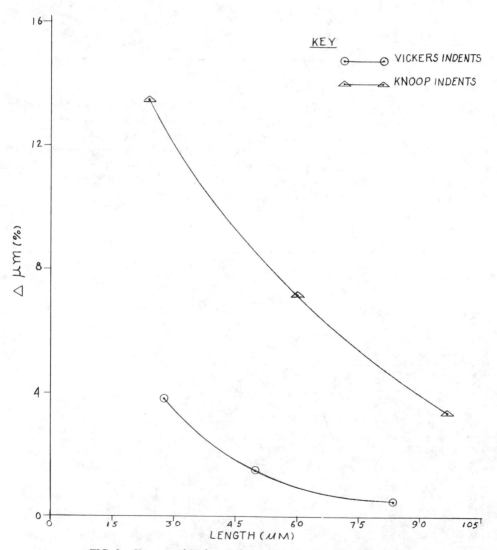

FIG. 2—*Knoop and Vickers indentations (data from Tables 3 and 4).*

TABLE 4—*Knoop indentation—steel 52100.*

Test No.	X_{cm}	X_{IA}	$X_{cm} - X_{IA}$, μm	HK_{cm}	HK_{IA}	$HK_{IA} - HK_{cm}$, HK	μm, %	HK, %
			30-G LOAD					
16S	23.4	20.5	2.9	780	1016	236	12.4	30.3
17S	23.8	20.5	3.3	754	1016	262	13.9	34.7
18S	23.5	20.7	2.8	773	996	223	11.9	28.8
19S	23.7	19.9	3.8	760	1078	318	16.0	41.8
20S	23.9	20.7	3.2	747	996	249	13.4	33.3
			Avg 3.20				13.52	33.78
			200-G LOAD					
21S	60.8	57.2	3.6	770	870	100	5.9	13.0
22S	60.6	54.5	6.1	775	958	183	10.1	23.6
23S	60.3	55.0	5.3	783	941	158	8.8	20.2
24S	60.1	57.2	2.9	788	870	82	4.8	10.4
25S	60.2	56.4	3.8	785	895	110	6.3	14.0
			Avg 4.34				7.18	16.24
			500-G LOAD					
26S	97.6	95.1	2.5	747	787	40	2.6	5.4
27S	97.1	93.1	4.0	755	821	66	4.1	8.7
28S	97.0	93.7	3.3	756	811	55	3.4	7.3
29S	97.3	94.3	3.0	752	800	48	3.1	6.4
30S	97.1	93.7	3.4	755	811	56	3.5	7.4
			Avg 3.24				3.24	7.04

on a conventional microscope. This bias measurement is somewhat proportional to the size of the indent, with the large indent approaching the true diagonal values.

The technique of using a known size indentation to calibrate the field, thus compensating for the lesser resolution and detective inadequacies, permits closer correlation to the conventional measurement method.

Measurements done on the Knoop and Vickers indentations, ranging from approximately 25 to 98 μm, show considerably more discrepancy with the Knoop than with the Vickers. Also, the Knoop errors are somewhat constantly between 3 to 4 μm regardless of the size of the indentation, while the Vickers indents have a proportional decrease in error with an increase in size.

Edward L. Tobolski[1]

Factors That Affect the Accuracy of Indentation Hardness Tests[2]

REFERENCE: Tobolski, E. L., **"Factors That Affect the Accuracy of Indentation Hardness Tests,"** *Factors That Affect the Precision of Mechanical Tests, ASTM STP 1025,* R. Papirno and H. C. Weiss, Eds., American Society for Testing and Materials, Philadelphia, 1989, pp. 46–51.

ABSTRACT: Rockwell, Brinell, and microhardness are the common indentation hardness tests discussed in this paper. Indentation hardness testing, in many cases, can be properly considered a nondestructive test since the part is usable after the test. The destructive effect at the test point, however, makes it impossible to retest the same point to verify the accuracy of the testing process. It is, therefore, critical that the first test be performed with a high degree of accuracy. The following items can seriously affect that accuracy: test material, testing instrument, operator, and environment.

Test Material

The nature of the material (grain structure, alloy, hardness, etc.), in many cases, will dictate the type of test required for best results. It is, therefore, important to be aware of the effectiveness of the different methods for the material being tested. The material itself should be of uniform hardness. Subsurface hardness variations can greatly alter the test result, masking the true value. This is also true if the part thickness is not appropriate for the test load applied. The test surface finish can have a similar effect on the hardness level as well as on repeatability. Powder metals present unique testing considerations since the hardness measured is not representative of the hardness of the constituent particles.

Testing Instrument

ASTM standards E 18[3], E 10[4], and E 384[5] precisely define the test parameters. The tester must be able to operate well within those limits. In addition, industrial practice requires that the tester be repeatable within a much narrower range than allowed by the ASTM standards. The integrity of the test loads, the rate of load application, and the measuring system are essential in the performance of an accurate test. The testing instrument must be capable of performing the same day to day, with minimum operator intervention. A proper monitoring program must be established to check the tester frequently. Test blocks are most commonly used for this purpose. Since test block hardness can change as the surface fills up with indents, care must be taken to maintain an effective program. Indenters are often manufactured from diamonds. The difficulty in producing identically shaped indenters from this hard substance can cause significant variations in hardness values, especially in the harder ranges.

---

[1] Vice president, Wilson Instruments, Binghamton, NY 13905.
[2] All hardness testing uses metric units for loads. Rockwell hardness numbers are unique, empirically derived values using a metric base dimension.
[3] ASTM Test Methods for Rockwell Hardness and Rockwell Superficial Hardness of Metallic Materials (E 18).
[4] ASTM Test Method for Brinell Hardness of Metallic Materials (E 10).
[5] ASTM Test Method for Microhardness of Materials (E 384).

Operator

Most modern testers have features to reduce the possibility of operator influence, particularly in optically measured hardness values. Equipment setup, fixturing, and test point locations are a few of the items that must be controlled if a proper testing program is to be implemented.

Environment

Hardness testers are high-precision instruments capable of measuring differences as small as a few microns. Dirt, dust, moisture, mechanical vibrations, electrical noise, and temperature variations can produce significant errors in the test results.

To establish a sound hardness testing program, all of the above items must be carefully considered if high degrees of accuracy and confidence are the desired result.

KEY WORDS: indentation hardness testing, indenters, grain structure, microhardness, case depth, decarburization, Rockwell test, superficial, dwell time, major load, minor load, diamond penetrators

All testing programs require a strict adherence to proper techniques to obtain accurate results. Hardness testing is no exception. The many factors that are a part of hardness testing make it very important that we are aware of the errors that could result. This paper could not possibly detail the magnitude of the effect of all the variables that are encountered; however, I will attempt to discuss as many items as possible and provide the reader with an indication of the possible impact each item could have on the test results. Most of my comments will be related to Rockwell testing since it is the most common test; however, many items will relate to Brinell and microhardness testing also.

Table 1 is a summary of the factors that affect test results. Each factor is discussed in detail below.

Test Selection

To select the proper test, you must know something about the material: its grain structure, its approximate hardness, its physical configuration, and its internal properties.

The grain structure could dictate the type of test used. For example, cast irons generally require Brinell tests that cover a relatively large area. If you were to do a C scale Rockwell test on most cast irons, the results could vary widely (20 to 30 points) depending upon the constituent that the smaller Rockwell indent hits.

Surface finish could also define the test used. A rougher surface would require a heavier load and/or larger indenter. A Brinell test may be necessary since it would be less affected

TABLE 1—*Items that affect the accuracy of indentation hardness tests.*

Item	Potential Effect
Test selection	From 0.5 to 2.0 HRC
	Up to 20 to 30 HRC
Testing instrument	1.0 to 5.0 HRC
Test loads and applications	Up to 2.0 HRC
Indenters	0.5 to 2.0 HRC
Environment	0.1 to 10.0 HRC
Operator	All the above

by the surface than a Rockwell or microhardness test. A "C" scale Rockwell test done on a surface of 60 RMS (root mean square) could vary from 2 to 5 points depending upon the hardness range of the part. While Brinell tests are more tolerant of finish, there are limitations that must be considered, and, in general, uniformity of finish is important to obtain reproducible results.

Material thickness and depth of case hardening also present problems when selecting the proper test. Significant errors can be obtained (10 to 20 points) when the proper test is not used. Those errors can also be very small as you get closer to the correct test, so it is easy to be fooled into thinking your test is accurate when it is not. Consult the appropriate ASTM standards for your test to determine if you are within the minimum thickness parameters. Material too thin will yield harder results as a result of the anviling effect. Note that, if you do testing on material too thin to support the test load, you risk the possibility of damaging the indenter, which could compromise the accuracy of other testing done with that indenter. Decarburization of case hardness conditions can also cause significant errors (as much as 20 points). In these cases, surface preparation and process knowledge become critical. Power metals present unique testing problems since the hardness measured is not representative of the hardness of the constituent particles, but rather a combination with the matrix. The densification of the material will affect both hardness levels and uniformity of results. Variations up to 4 to 5 HRC are not unusual.

Testing Instrument

The design, assembly, and condition of the testing apparatus are all critical to accurate testing. Excessive friction at any one of several areas can cause errors in both level and repeatability. Even testers that have no obvious operational problems can give poor results of 1 to 2 points due to excessive friction in the loading system, usually on the hard side. Similar errors can be expected from the measuring system due to small amounts of frictions. This usually gives results too soft.

Excessive deflections of the supporting frame of the tester and the test piece support system can cause problems also. Errors of 1 to 3 points are not uncommon due to improper anviling of the parts and excessive deflection of the tester's frame.

The loading mechanism must provide for constantly accurate loads applied properly, well within ASTM standards. The effects of error in this area are discussed in the following section.

The measuring system is critical considering the very small dimensions being measured. One regular scale Rockwell point is 80 millionths of an inch (2 μm) and the superficial scale is half that; therefore, measuring system accuracy is very important. Analog gages were popular for many years; however, they were subject to mechanical errors up to 0.5 point. This, combined with the interpretation and bias error of the operator, could result in up to ± 1.0 point error. Modern, fully digital, electronic systems are much superior, with encoder errors less than 0.1 point with no operator interpretation.

Test Loads and Application

The load parameters of the various tests are identified in the appropriate ASTM standards. High-quality testing equipment should be able to apply loads well within those requirements. Figures 1 and 2 give an indication of the errors as a result of a 10% error in the minor load. The effect is related to the material hardness, with softer materials having larger errors. Figure 2 shows the effect of a 0.6% error in the major load. It should be noted

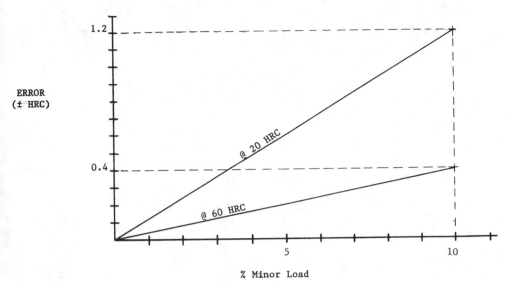

FIG. 1—*Effect of error in minor load.*

that the magnitude of the error in both figures is the same (1 kg); the percent error changes due to the load differences.

The application of the test loads involves both the velocity and the dwell time of the major loads. Variations of application velocity that can be obtained with some manually controlled testing devices can give test result variations up to 1.0 HRC at HRC 60. Softer materials and materials subject to work hardening could give significantly higher errors. Major load dwell times are specified in ASTM E 18. Longer dwell times will give errors up to 0.6 points as indicated in Fig. 3. Other hardness scales, such as B scale could give significantly larger errors.

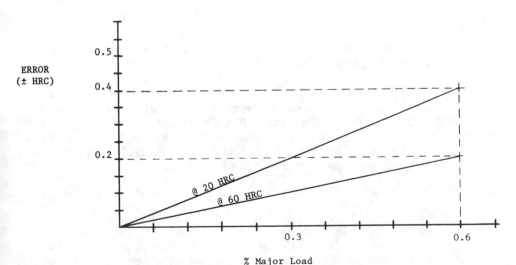

FIG. 2—*Effect of error in major load.*

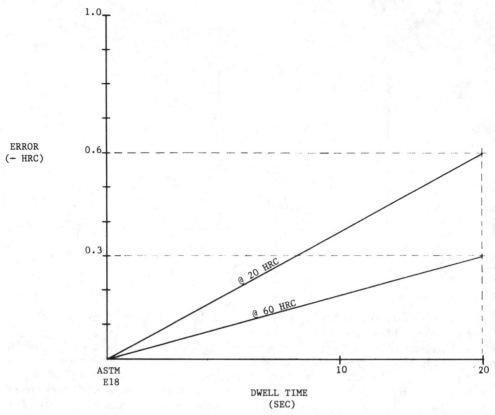

FIG. 3—*Effect of major load dwell time.*

Indenters

The indenter, being a part of the measuring system, is critical to the accuracy of the test. Ball indenters are less a problem because it is relatively easy to make a ball penetrator to the required tolerances, and the round geometry eliminates some problems. Diamond penetrators are, on the other hand, very difficult to manufacture. The potential sources of error are significant; I will not attempt to categorize the effect of each. It is important to note, for this discussion, that the best production diamond penetrators made today will exhibit variation up to 0.5 HRC when compared on the same tester. Lesser quality penetrators will give significantly larger errors. This indenter error is significant because, unlike all other factors, it is impossible to eliminate completely.

Environment

Temperature and humidity within the normal range of working conditions should not affect test results significantly (less than 0.1 HRC). Operations outside normal ranges should be cause for concern. Comparative testing should be done to isolate the effect.

Cleanliness, vibrations, and electrical interference can cause significant problems that are difficult to qualify. Ultra low load (1 g) microhardness testing requires an absolutely vibration-free environment, whereas tests above 25 g are not so critical.

Operator

The best testing program can be compromised if the operator is not properly trained and motivated to do a good job. Test locations become important on many samples. Tests near the edge of a part or near each other must be properly located to insure an accurate test. Errors of up to 2 HRC are not uncommon without any obvious indication. Overall monitoring of the operation is very important. Some modern testers have features to minimize operator error, but the person remains the vital link in the successful testing program.

Bibliography

Petik, F., "Factors Influencing Hardness Measuring (A Systematic Survey of Research)," circulated by OIML, Paris, France, 1981.

Joseph J. Cieplak,[1] Edward L. Tobolski,[2] and Dennis M. Williams[2]

Gage Repeatability and Reproducibility Studies of Rockwell Scale Hardness Testers

REFERENCE: Cieplak, J. J., Tobolski, E. L., and Williams, D. M., **"Gage Repeatability and Reproducibility Studies of Rockwell Scale Hardness Testers,"** *Factors That Affect the Precision of Mechanical Tests, ASTM STP 1025,* R. Papirno and H. C. Weiss, Eds., American Society for Testing and Materials, Philadelphia, 1989, pp. 52–60.

ABSTRACT: It is difficult to interpret the results of gage repeatability and reproducibility (GRR) studies on Rockwell scale test instruments because (1) the test can never be performed in the same place more than once, (2) no material is completely uniform in hardness, and (3) the expected variation in hardness is itself hardness dependent.

Therefore, the concept of material variation (MV) must be added to the already recognized sources of variation called equipment variation (EV) and appraiser variation (AV).

Following specific procedures outlined in this paper will insure performance of the most realistic GRR study. Interpretation of the results using methods suggested in this paper will aid in understanding the information gathered in this study.

KEY WORDS: hardness testing, Rockwell, gage repeatability and reproducibility (GRR), statistical process control, hardness standards

Statistical process control (SPC) methods have become an increasingly important tool to insure high quality products at a cost-effective price. The gage repeatability and reproducibility (GRR) study is one of the elements in a total SPC program which determines the useful working range of the inspection equipment being used.

It is difficult to interpret the results of GRR studies on Rockwell scale test instruments because of factors specific to the Rockwell test. This paper discusses those factors and also recommends procedures to follow which will minimize misunderstanding and maximize the precision of the hardness testing data. SPC methods were first used by American industry in World War II, but their importance was not recognized until the late 1970s. Now SPC is reversed as the new "gospel" of American industry. Proper implementation of SPC methods will cut costs to the minimum, provide a high quality product, and pinpoint and prioritize future direction.

Inspection equipment used in quality control programs is itself subject to variations in performance which may be operator induced, equipment induced, procedurally induced, or environmentally induced.

The GRR study is a procedure for determining the repeatability (precision) of a test instrument and the reproducibility (variation among operators) of a specific gage in operation.

In brief, a GRR study collects measurements made by a number of operators on a num-

[1] Vice president—Marketing, Bryce Office Systems, Inc., Oxford, CT 06483-1011.

[2] Vice president—Manufacturing, and applications engineer, respectively, Wilson Instruments, Binghamton, NY 13905-2508.

ber of specimens at different times, or trials, but all taken on the same test instrument (gage). A mathematical analysis of these results then provides values for equipment variation—(repeatability), appraiser variation (reproducibility), and an overall R&R index.

The Rockwell hardness test provides a relative value of the hardness of the material being tested. This value is then correlated to other characteristics, such as strength and wearability. To perform the test, a calibrated indenter, either a diamond cone or a ball, is forced into the specimen under a fixed "minor" load to establish an initial reference point. The full "major" load is then applied and removed. The final Rockwell result is determined by measuring the resulting deformation between the initial reference point and the final recovered depth of the indentation. The Rockwell test instrument is a very precise piece of equipment. One Rockwell point is equivalent to 2.0 μm (approximately 80 millionths of an inch) in depth.

The GRR Study

A variety of methods for performing the GRR study are in use throughout the United States. All methods are directed toward determining the inherent variability of a gage in use monitoring a specific process. An acceptable gage is one which has an inherent variability of 10% or less of the process tolerance. Gages with an inherent variability ranging from 10 to 30% of process tolerance may be acceptable for use on an interim basis. Gages with an inherent variability greater than 30% are not acceptable for use [1].

We have found the methods and procedures outlined in section 3.4 of the *General Motors Statistical Process Control Manual* published by the SPEAR administrative staff at the GM Tech Center to be the most representative and well documented for general use. While other methods may work perfectly well, the GM procedures are logically formatted, complete with sample forms, straightforward to use, and gaining widespread use because of GM's ability to require vendor participation. Section 3.4 covers the "Variable Gage Study for Repeatability and Reproducibility (Long Method)."

Under this procedure, ten specimens are tested in three separate trials by three different operators for a total of 90 tests, 9 on each specimen. It is important that the specimens be tested in statistically random order by each operator in each trial to insure statistically valid results.

The GM SPEAR guidelines provide a data collection form (Fig. 1) on which to record the data as it is taken by each operator.

Once the testing is completed, the range of readings taken by each operator is calculated, and the average range, $\overline{R}a$, $\overline{R}b$, $\overline{R}c$, is calculated for each operator. Then the average range of all three operators is calculated, $\overline{R}$.

Also calculated are the average test values determined by each operator, $\overline{X}a$, $\overline{X}b$, $\overline{X}c$. The minimum average test value, MIN.$\overline{X}$, is subtracted from the maximum average test value, MAX.$\overline{X}$, to determine the difference between the average test values, $\overline{X}$ Diff.

The complete analysis for the final repeatability and reproducibility index (R&R index) is calculated using the worksheet and formulas shown in Sections 3.4 (Fig. 2).

The repeatability, or equipment variation (EV), is the variance within numerous sets of readings received by several operators from a single piece of equipment. This represents the variation a typical operator can expect to get from the equipment.

$$EV = (\overline{R}) \times (K_1)$$

where $K_1 = 3.05$ for three trials.

The % EV is used to show the amount of equipment variation as a percentage of the process/part tolerance.

GAGE REPEATABILITY AND REPRODUCIBILITY DATA SHEET (Long Method)

Operator	A – Ed Tobolski				B – Joe Cierplak				C – Linda Mahoney			
Sample #	1st Trial	2nd Trial	3rd Trial	Range	1st Trial	2nd Trial	3rd Trial	Range	1st Trial	2nd Trial	3rd Trial	Range
	1	2	3	4	5	6	7	8	9	10	11	12
1	45.6	45.7	45.8	0.2	45.8	45.8	45.9	0.1	45.5	45.5	45.6	0.0
2	44.3	44.2	44.0	0.3	44.4	44.4	44.5	0.1	44.5	44.8	44.6	0.3
3	45.4	45.3	45.4	0.1	45.3	45.4	45.3	0.1	45.3	45.4	45.3	0.1
4	45.6	45.7	45.6	0.1	45.8	45.8	45.8	0.0	45.6	45.7	45.8	0.2
5	45.1	45.0	44.9	0.2	44.9	44.9	45.2	0.3	45.2	45.0	46.0	0.2
6	45.7	45.7	45.8	0.1	45.7	45.7	45.6	0.1	45.6	45.7	45.4	0.3
7	45.7	45.9	45.6	0.3	45.8	45.8	45.9	0.1	45.8	45.8	45.7	0.1
8	46.7	46.5	46.4	0.3	46.5	46.3	46.3	0.2	46.7	46.5	46.5	0.2
9	46.6	46.7	46.7	0.1	46.8	46.8	46.6	0.2	46.7	46.8	46.6	0.2
10	46.2	46.2	46.4	0.2	46.3	46.3	46.4	0.1	46.5	46.4	46.4	0.1
Totals	456.9	456.9	452.6	1.9	457.3	457.2	457.5	1.3	457.5	457.6	456.8	1.7

Operator A: Sum 456.6, $\bar{R}_A$ 0.19; Sum 1370.4, $\bar{X}_A$ 45.68

Operator B: Sum 457.3, 457.5, $\bar{R}_B$ 0.13; Sum 1372.0, $\bar{X}_B$ 45.73

Operator C: Sum 457.5, 456.8, $\bar{R}_C$ 0.17; Sum 1371.9, $\bar{X}_C$ 45.73

# Trials	D_4
2	3.27
3	2.58

$\bar{R}_A$	.19
$\bar{R}_B$	.13
$\bar{R}_C$	.17
Sum	.49
$\bar{R}$	.163

Max. $\bar{X}$	45.73
Min. $\bar{X}$	45.61
$\bar{X}$ Diff.	0.05

$(\bar{R}) \times (D_4) = UCL_{R^*}$

$(.163) \times (2.58) = .42$

*Limit of individual R's. Circle those that are beyond this limit. Identify the cause and correct. Repeat these readings using the same appraiser and unit as originally used or discard values and reaverage and recompute R and the limiting value UCL_R from the remaining observations

NOTES: Wilson®/Rockwell® Model 553T Hardness Tester SN-09198660 4, 12/17/96 — HRC scale — Delete first test to seat sample — Ten test blocks, HRC 44–47 Nominal Value.

FIG. 1—GM SPEAR gage repeatability and reproducibility data sheet (long method) with data.

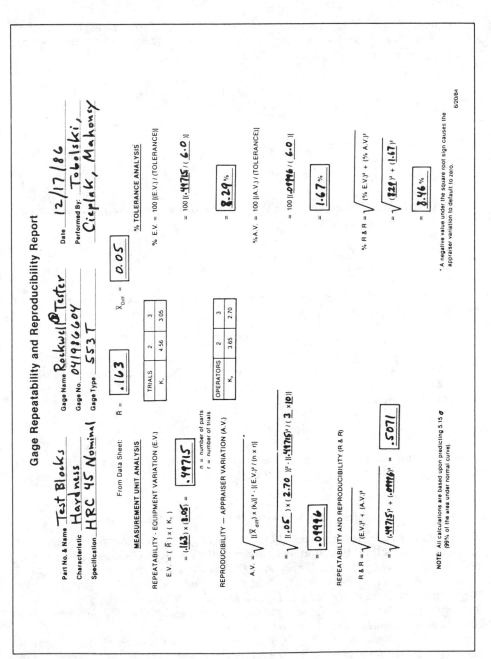

FIG. 2—*GM SPEAR gage repeatability and reproducibility report with data.*

$$\% \text{ EV} = 100 \times \frac{(\text{EV})}{(\text{Tolerance})}$$

The reproducibility, or appraiser variation (AV), is the variation of average readings obtained by all operators from a single piece of equipment. This represents the bias that can be expected between operators.

$$\text{AV} = [(\overline{X} \text{ Diff.}) \times (K_2)]^2 \cdot \left[\frac{(\text{EV})^2}{(n \times r)} \right]$$

where

K_2 = 2.70 for three operators,
n = number of parts, and
r = number of trials.

The % AV is then calculated to produce the amount of appraiser variation as a percentage of the process/part tolerance.

$$\% \text{ AV} = 100 \times \frac{(\text{AV})}{(\text{Tolerance})}$$

Finally, the overall R&R index is determined by taking the square root of the sum of the squares for EV and AV

$$\text{R\&R} = \sqrt{(\text{EV})^2 + (\text{AV})^2}$$

The resulting value for R&R is the variance that can be expected from a measuring device that is due to the (operator) use of that same device. The R&R is then shown as a percentage of the tolerance.

$$\% \text{ R\&R} = (\% \text{ EV})^2 + (\% \text{ AV})^2$$

You can readily see that the tolerance assigned to the process has as much to do with the percentage calculations as does the performance of the equipment and the appraiser.

Test Specific Factors Which Make Rockwell GRR Studies Difficult to Interpret

By its nature as a physical mechanical test, the Rockwell test has specific characteristics which make it difficult for investigators to interpret the results of GRR studies.

First, you can never perform another test in the same location once you have performed the first test. When the test is made, the material around and beneath the indent is permanently deformed and its hardness changed by the test itself. The material in this area is essentially work hardened. Subsequent measurements taken at that site, or within three diameters of the indent from that site, would be invalid.

A second test-specific characteristic is that no materials are completely homogeneous. Even a perfect test instrument would never get completely consistent hardness values because the material's hardness varies from test site to test site.

A third specific characteristic of the hardness test is that the expected variation of the test results are hardness and scale dependent. This means that, as you test softer and softer materials, you should expect to get greater and greater variation in the uniformity of the hardness readings obtained. This is a consequence of the metallurgical properties of the specific material being tested.

The importance of this fact to GRR studies is that they should be done at the hardness range you expect from your production parts, and if you use that test instrument at a different hardness range, then you should expect to obtain different GRR results.

These three factors cause us to introduce the concept of material variation (MV) into the evaluation of the GRR index. So, for a hardness test, we now have three sources of inherent variability:

1. EV = equipment variation,
2. AV = appraiser variation (operator), and
3. MV = material variation.

If you examine the data collection and analysis worksheets shown in Figs. 1 and 2, you can see that the larger the average range (R), the larger the EV, and to a smaller extent, the larger the AV. Likewise, larger variation in the average difference of the test values ($\overline{X}$ Diff.) will cause a larger AV.

When you understand that there is a material variation element in the results of the hardness test, you must also recognize that this element contributes to a falsely larger GRR index than is truly present if the inherent variability of the material could be factored out. Since this variability cannot be factored out, it will, in turn, produce a higher value for the GRR index. It is important to note, however, that material variation is not present in actual testing because material hardness is the variation being measured.

Can the inherent variability of the material be factored out? Not at present. There is a theoretical method of removing the inherent variability of the material. This would be to perform the test on a hard material with a blunt indenter. This will give uniform results; however, they will be infinitely hard. Remember that the GRR study must be performed at the approximate hardness range where your production will be run, so this is not a valid alternative.

So what does the investigator do to minimize the influence of the MV factor? First, he performs the GRR study using standardized test blocks as his specimens, not production parts. This will significantly reduce the MV factor.

Test blocks are inherently more uniform than production parts, thereby reducing the influence of the material variation in the results. Secondly, you must be willing to accept a higher R&R index than the 10%/30% rule would dictate because of the presence of the MV factor in the GRR study, which is not present in actual production testing.

A visual interpretation of the presence and consequences of the three causes of inherent variability is shown in Fig. 3. Notice that the EV factor is generally uniform throughout the working hardness range of the equipment, while the AV component increases, with softer materials contributing in an increasingly significant way to the overall R&R index.

The example shown is based upon a real-life study performed on a Wilson/Rockwell Model 553T hardness tester on HRC 45 test blocks. The R&R index calculated in this study was 0.5071, as shown in Fig. 2.

As stated earlier, the specified process tolerance directly affects the percent R&R index.

Recommended Procedures for Conducting the Rockwell Test GRR Study

1. Follow the General Motors SPEAR guidelines for the long method using three operators and three trials. This will give you the most consistent results. While there are other

Hardness

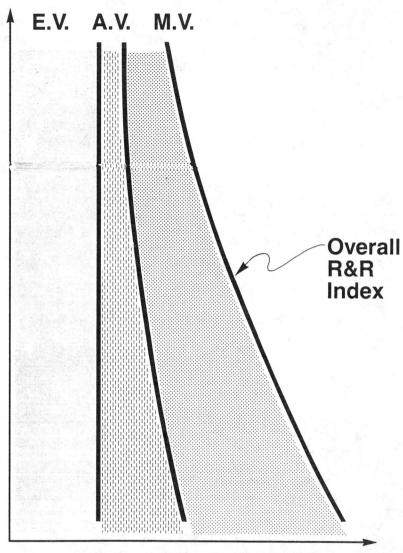

FIG. 3—*The three components of inherent variability: EV, AV, and MV.*

methods which will provide acceptable results, the GM SPEAR procedure is well documented, explained in good detail, and in general use throughout most of industry.

2. Use standardized test blocks, not production parts. Test blocks should have a nominal hardness value within 10 Rockwell points of the target value for your process.

3. Seat the anvil and penetrator with a "throw-away" seating test before beginning to collect actual test data. This is normal hardness testing practice and helps insure more uniform results.

4. Do not change the penetrator or anvil during the GRR study. Different penetrators may add a one-half point bias to the results obtained. While this type of bias is consistently repeatable within a GRR study, a change during the procedure can introduce another source of variability.

5. Use the same anvil as you would with production parts. Standard procedure calls for using a pedestal spot anvil when performing a calibration check on standardized test blocks. But, when conducting a GRR study, you should use the production anvil because it is a part of the testing system when actual parts are checked. This will give you a more realistic appraisal of the testing system's performance in actual practice.

6. Make certain that testing is done in random order to preserve the statistical integrity of the test.

7. Remember to follow good, accepted testing procedures. The tester should be in good working order, recently certified for proper calibration and condition. The minor and major loads should be applied smoothly and consistently at the correct rate of load application. The dwell and recovery times must be in accordance with the standard. The gage must be read carefully and correctly. It is important to understand that while the GRR study examines repeatability and reproducibility, it does not at all address the accuracy of the test results.

It does little good to obtain consistent test results if they are all consistently five points lower than what the true value should be. Therefore, it is critical that the accuracy level of the test equipment be verified with standardized test blocks in accordance with ASTM Test Methods for Rockwell Hardness and Rockwell Superficial Hardness of Metallic Materials (E 18-84) before running the GRR study.

How to Qualify Your Hardness Tester as an Acceptable Gage

If you have performed your GRR study in accordance with the previous suggestions and your tester qualifies with an R&R index of less than 10% of process tolerance, congratulations.

If your tester still does not qualify, here is a list of suggestions which will improve your tester's ability to meet or exceed the required levels of performance.

Follow these suggestions in whatever order makes sense to you and your organization in terms of cost and appropriateness.

1. *Better operator training.* This will insure more reproducible results. Operators must be trained to read the test results correctly. On a dial gage (analog) tester, the dial needle must be read when it is directly perpendicular to the dial face. The operator must estimate the reading to the nearest tenth of a Rockwell point. Additionally, both the minor load and major load must be applied smoothly and consistently with no impact. Also, the dwell and recovery times should be consistent and in agreement with the standard. Operator training is generally the most effective method of reducing the R&R index.

2. *More uniform test blocks.* Not all standardized test blocks have the same amount of inherent variability. ASTM recognizes this in Standard E 18-84 for Rockwell hardness testing when it says that on an HRC 63 test block the acceptable variation is 1.0 points, and at HRC 25 the acceptable variation is 2.0 points. Therefore, it is understood that although all test blocks are quite uniform in hardness, some are more uniform than others. If this item is critical, then you should sort through your standardized test blocks and use only those with the greatest uniformity for your GRR study.

3. *Use a clamping device.* Do this to steady the part being tested and to reduce any variations possibly caused by movement of the part during testing.

4. *Repair the tester or have it serviced.* This will bring the particular tester up to its best operating condition. Often this is the single best course of action to follow in conjunction with better operator training.

5. *Replace the tester with a model with less inherent variability.* If the tester is worn and certified to be in fair or poor condition, then replacement may be the only reasonable alternative.

6. *Replace the tester with a model which minimizes operator influence on the test results.* A digital readout model is inherently more repeatable than a dial gage model because the readout device is more accurate and easier to read reliably. A tester with automatic or motorized application of the minor and major loads is inherently more repeatable than a manual model where different operators can obtain variable results. The more the test is controlled by the tester and the less it is controlled by the operator, the lower will be the R&R index and the greater your chances of qualifying the gage.

7. *Change process tolerance requirements.* Increasing the acceptable process tolerance will have a direct effect on decreasing your percent R&R index. Some processes may have unnecessarily tight tolerance requirements on them which could be modified with no detriment to the product.

Expected Performance by Equipment Type

An analog gage type tester that offers a manual loading system produces the poorest GRR indexes. This is because operator error is introduced when the load is applied and when the resulting hardness is read from the dial.

On the other hand, automatic testers that provide an automatic loading system and digital readout have been found to lower GRR indexes considerably. This is a result of minimizing operator influence (AV).

Reference

[1] *General Motors Statistical Process Control Manual,* SPEAR Administrative Staff, GM Tech Center, 1986, pp. 3–10.

Bibliography

DataMyte Handbook, 3rd ed., DataMyte Corp., Minnetonka, MN, June 1987.

Fatigue and Fracture Procedures

Masanobu Satoh,[1] Tatsuo Funada,[1] Yoshio Urabe,[1] and Kiminobu Hojo[1]

Measurement of Rapid-Loading Fracture Toughness J_{Id}

REFERENCE: Satoh, M., Funada, T., Urabe, Y., and Hojo, K., **"Measurement of Rapid-Loading Fracture Toughness J_{Id},"** *Factors That Affect the Precision of Mechanical Tests, ASTM STP 1025,* R. Papirno and H.C. Weiss, Eds., American Society for Testing and Materials, Philadelphia, 1989, pp. 63–76.

ABSTRACT: The effect of the loading rate on elastic-plastic fracture toughness, which defines stable crack initiation, has not been clear; therefore, the test method of elastic-plastic fracture toughness under rapid loading (rapid-loading fracture toughness J_{Id}) has not yet been established in ASTM Test Method for J_{Ic}, a Measure of Fracture Toughness (E 813-81).

The authors studied the J_{Id} test method by using a high-speed servohydraulic testing machine, by which the rapid-loading fracture toughness test could be done in accordance with ASTM Test Method for Plane-Strain Fracture Toughness of Metallic Materials (E 399-83 ANNEX A7). We evaluated the applicability of the unloading compliance method to determine the stable crack initiation under rapid loading by comparing the results with the multiple specimen method.

We also applied the electric potential method, which is expected to be able to detect crack growth continuously to determine the stable crack initiation. Results are as follows.

1. The unloading compliance method is applicable to rapid-loading tests. J_{Id} measured by the unloading compliance method is the same as that obtained by multiple specimen method.

2. J_{Id} based on the electric potential method is about the same as that of the unloading compliance method.

3. J_{Id} depends on loading rate. Transition temperature between the brittle-ductile region tends to be higher as the loading rate increases.

KEY WORDS: rapid-loading fracture toughness, elastic-plastic fracture toughness, unloading compliance method, multiple specimen method, electric potential method, high-speed servohydraulic testing machine, A 533 steel, A 508 steel

It is well known that for many materials such as carbon steels, as the loading rate increases, the plane-strain fracture toughness, K_{Ic}, decreases and the brittle-to-ductile transition temperature increases. Considering these matters, the fracture toughness test method under rapid loading is specified in ASTM E 399-83 ANNEX A7.

The effect of the loading rate on elastic-plastic fracture toughness, J_{Ic}, which defines stable crack initiation, has not been clear. Currently, there is no standard test method for determining J_{Ic} under rapid loading (rapid-loading fracture toughness J_{Id}) in ASTM E 813-81.

The authors studied the J_{Id} test method by using a high-speed servohydraulic testing machine, by which the rapid-loading fracture toughness test could be done in accordance

[1] Assistant chief research engineers and research engineer, respectively, Mitsubishi Heavy Industries, Ltd., Takasago Research & Development Center, Takasago 676, Japan.

with ASTM E 399-83 ANNEX A7. We evaluated the applicability of the unloading compliance method to determine the stable crack initiation even under rapid loading by comparison with the results from the multiple specimen method.

We also applied the electric potential method to determine the stable crack initiation. This method is also expected to be able to detect crack growth continuously.

Materials

The materials used in the test program were: ASTM Specification for Pressure Vessel Plates, Alloy Steel, Quenched and Tempered, Manganese-Molybdenum and Manganese-Molybdenum-Nickel (A 533) Grade B Class 1 (A 533 Gr.B Cl.1); ASTM Specification for Quenched and Tempered Vacuum-Treated Carbon and Alloy Steel Forgings for Pressure Vessels (A 508) Class 3 (A 508 Cl.3); and A 533 Grade B Class 2 (A 533 Gr.B Cl.2).

A 533 Gr.B Cl.1 and A 533 Gr.B Cl.2 are typical commercial plates for nuclear pressure vessel components, and A 508 Cl.3 is also a typical commercial forging for nuclear pressure vessels.

Chemical compositions and mechanical properties of the materials are presented in Table 1.

Experimental Procedure

Specimen Preparation

The 1T C(T), a typical compact specimen modified for J-integral testing specified by ASTM E 813-81, was used for the majority of specimens. In the case of A 533 Gr.B Cl.1, a disk-shaped compact specimen, 1/2T DC(T), of one-half size, was also used to determine the effect of small-size specimens.

The preparation of fatigue cracks was subjected to ASTM E 399-83 ANNEX A2; the final value of a/w was 0.55.

The specimens were side grooved to a depth of 12.5% per side. The angle of the side groove was 45° and the root radius was 0.25 mm.

TABLE 1 Chemical compositions and mechanical properties of the materials used.

Material	Thickness mm	Chemical Composition, weight %									Tensile Properties			Charpy V-notch Impact Properties		NDT Temperature °C
		C	Si	Mn	P	S	Cu	Ni	Cr	Mo	σ_{ys} MPa	σ_{ut} MPa	Eℓ. GL=50 %	FATT °C	USE J	
A533 Gr.B Cℓ. 1	200	0.22	0.28	1.26	0.012	0.016	0.18	0.54	0.11	0.48	422	621	26.8	20	80	−15
A508 Cℓ. 3	200	0.21	0.25	1.39	0.004	0.008	0.04	0.68	0.10	0.49	461	612	22.7	−11	181	−25
A533 Gr.B Cℓ. 2	100	0.19	0.22	1.44	0.007	0.001	0.01	0.65	0.16	0.54	552	680	26.8	−61	260	−50

Testing Apparatus

The MTS high-speed servohydraulic testing machine was used for the test, by which a rapid-loading fracture toughness test could be performed in accordance with ASTM E 399-83 ANNEX A7. Its maximum stroke speed is 500 mm/s, with a maximum loading capacity of 400 kN.

The load was measured using a load cell with a resonant frequency of 3.3 kHz. The load point crack opening displacement (COD) was measured using a COD gauge with a resonant frequency of 6.2 kHz when attached to the specimen. These transducers satisfy the requirements of ASTM E 399-83 ANNEX A7, completely applicable to rapid-load testing.

The high-rate signal amplifiers used for measuring load and COD have a high-frequency response from dc to 100 kHz and also satisfy the requirement of $20/t$ (kHz), ($t \geqq 1$ ms) as specified in ASTM E 399-83 ANNEX A7.

The signals from the load cell and COD gage were fed into transient memory (12 bit $\times$ 4000 point/channel with a maximum sampling rate of 1 μs/point).

The load-time and COD-time data were transferred from the transient memories of a high-speed oscilloscope to the microcomputer for analysis.

Procedure

Unloading Compliance Method

The dynamic fracture toughness in high loading rate (herein called rapid loading fracture toughness J_{ld}) was measured by both the multiple specimen method and the unloading compliance method by referring to ASTM E 813-81, which specifies the method for measuring the quasistatic loading fracture toughness J_{lc} for ductile cracks.

The unloading compliance method is specified in ASTM E 813-81 ANNEX A1, which, however, does not specify a case with a high loading speed. In the test of this report, the loading speed was set to a stroke speed of 500 mm/s, while the unloading was performed at a slow speed, such as the static test described ANNEX A1. Crack extension was calculated after transferring the relevant data from the crack extension analysis amplifier to the microcomputer to amplify the loads and COD data.

The J_{ld} measurement system, based on the unloading compliance method, is shown in Fig. 1.

Multiple Specimen Method

The multiple specimen method used in this study was based on ASTM E 813-81 at a loading speed of 500 mm/s.

Electric Potential Method

For the test of this report, the d-c electric potential measuring system as shown in Fig. 2 was devised and used for the measurement. Data obtained in the electric potential method were processed as follows.

The electric potential value of the dummy specimen was subtracted from the potential value of the specimen and smoothed by second and third order polynomials. Next, the data were combined with the COD versus time, creating the potential-COD curve, in which the crack initiation point was defined by the inflection point of the curve.

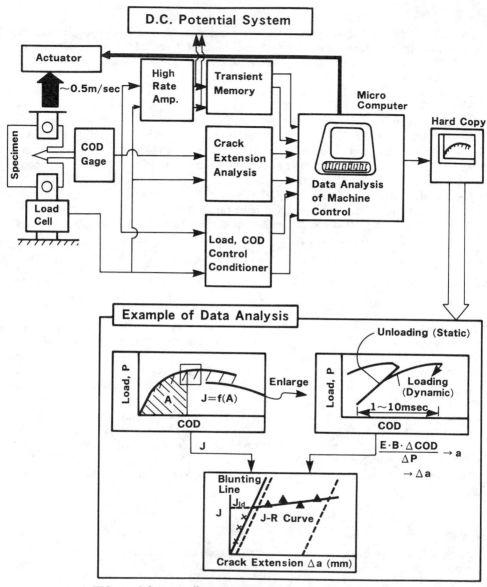

FIG. 1—*Schematic illustration of the* $J/_{Id}$ measurement system.

Results and Discussion

Crack Extension

The ASTM E 813-81 ANNEX A1 requires that the total amount of crack extension predicted by the unloading compliance method agree with the averaged value determined by the heat tint method within 15%. Figure 3 compares the predicted crack extension obtained by the unloading compliance method with the crack extension measured by the heat tint

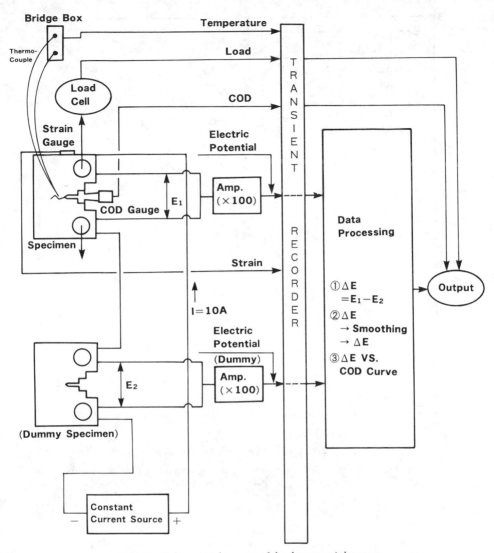

FIG. 2—*Schematic diagram of the d-c potential system.*

method. The two methods agree within ±15% in more than 80% of the cases shown in Fig. 3.

Accordingly, it was confirmed that the unloading compliance method used in this study was effective. Crack length, a_i, at the i-th partial unloading was calculated from Eq 1 [1] by using the unloading compliance, λ_i, and crack extension, Δa_i, was calculated from Eq 2. The heat tint method was performed exactly according to ASTM E 813-81.

$$a_i = W \cdot (1.002 - 4.0632U_i + 11.242U_i^2 - 106.04U_i^3 + 464.33U_i^4 - 650.68U_i^5) \quad (1)$$

$$\Delta a_i = a_i - a_o \quad (2)$$

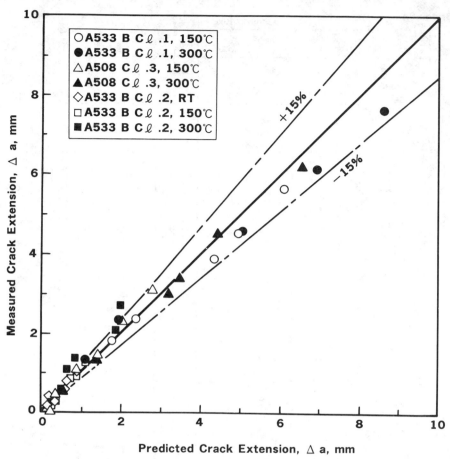

FIG. 3—*Comparison of the amount of crack extension predicted by the unloading compliance method with the measured value as determined by the heat tint method.*

where

$$U_i = \frac{1}{\sqrt{B_e E \lambda_i} + 1},$$

λ_i = compliance with i-th partial unloading after being corrected by angle [$\lambda_i = (COD/P_i)$],

P_i = load,

COD = load point crack opening displacement,

B_e = effective thickness of specimen, $B_e = B - [(B - B_N)^2/B]$,

B = thickness of specimen,

B_N = thickness at minimum section of the specimen with side groove,

E = Young's modulus,

a_o = initial crack length, and

W = specimen width.

Comparison of the Unloading Compliance Method with the Multiple Specimen Method

J_{Id} values, measured by the unloading compliance method and the multiple specimen method, are compared in Fig. 4. Seven out of ten tests showed agreement within $\pm 20\%$ for the two methods, indicating that the unloading compliance method is acceptable for measuring J_{Id} at a high loading rate. Figure 4 also shows the result of a 1/2T DC(T) test, a small specimen in which J_{Id} based on both methods agree satisfactorily with each other. In addition, the result of the 1/2T DC(T) obviously brings about substantially the same J_{Id} as a 1T C(T). Consequently, it is revealed that adoption of smaller specimens, that is by use of the 1/2DC(T) specimen, is effectively applicable to simplifying the test method.

Figures 5 and 6 show typical load-COD curves and J-R curves based on both methods. Figure 7 shows the relationship between Δa and COD. Referring to these figures, the difference between the load-COD curves obtained by the two methods is small, as long as COD is no greater than about 1.5 mm. However, as COD increases, the difference between these methods becomes larger, that is, significant load drop was observed in the case of the unloading method and there is a discrepancy between Δa obtained by both methods as a general tendency.

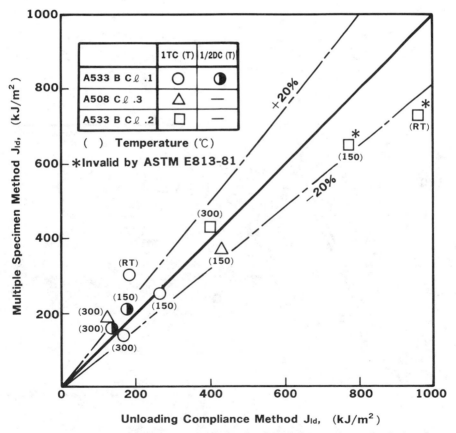

FIG. 4—*Comparison of* $J/_{Id}$ *obtained by the unloading compliance method with* $J/_{Id}$ *obtained by the multiple specimen method.*

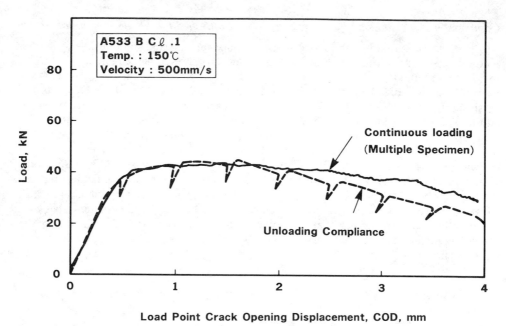

FIG. 5—*Typical load-versus-COD curve recorded on the unloading compliance method and the continuous loading (multiple specimen) method.*

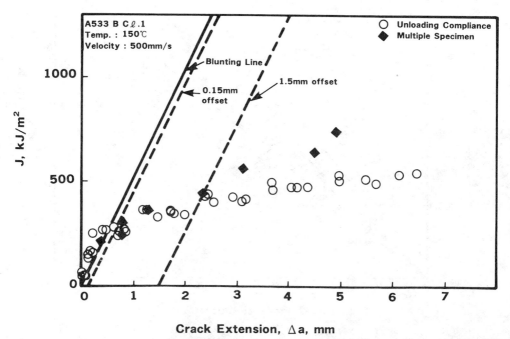

FIG. 6—*Comparison of J-R curve obtained by the unloading compliance method with that by multiple specimen method (A 533 Gr.B Cl.1).*

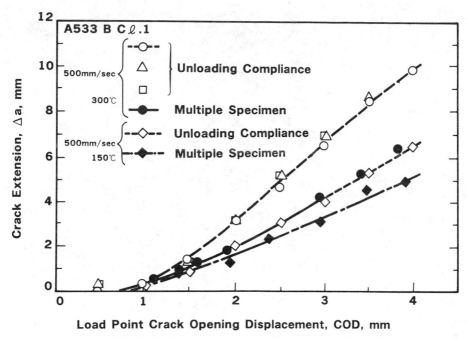

FIG. 7—*Relation between crack extension and COD.*

This trend is also observed in a typical crack extension photograph shown in Fig. 8. Relating to the comparison of the J-R curve, both methods give substantially the same results with the crack extension Δa up to approximately 2 mm. As Δa increases, the difference between Δa obtained by both methods becomes larger in general. J values obtained by the unloading compliance method are obviously smaller than those obtained by the multiple specimen method. These trends are more significant at higher temperatures.

In determining J_{1d}, both methods are applicable provided that Δa remain small.

When the dynamic fracture analysis is applied by using the J-R curve, however, further study is needed to determine if the J-R curve method is more suitable for evaluating the fracture of actual structure. The J-R curve obtained by the unloading compliance method might underestimate the actual fracture resistance as determined by the J-R curve obtained by the multiple specimen method.

Crack Initiation Detected by Electric Potential Method

A measurement by the d-c electric potential method was attempted to detect the crack initiation. The measuring system is already shown in Fig. 2.

Figure 9 summarizes the relationships between load–COD, electric potential–COD, and Δa–COD. Obviously, the crack extension phenomena are well detected by the increase of electrical potential. However, the COD value at crack initiation, measured by the electric potential method, is about the same as the COD value based on the unloading compliance method. Figure 10 shows a comparison of J_{1d} obtained by both methods, in which J_{1d} obtained by the electric potential method also brings about a slightly larger value.

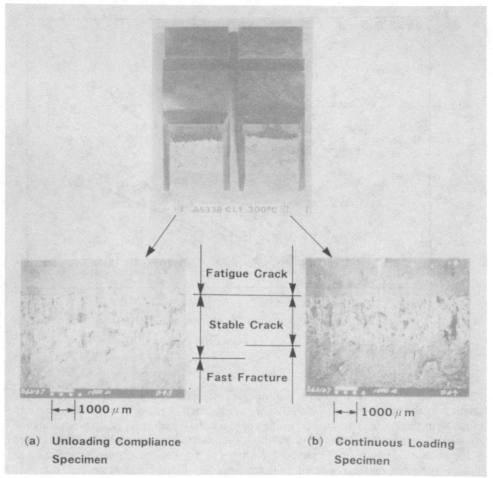

FIG. 8—*Examples of stable crack extension at 2-mm COD (A 533 Gr.B Cl.1).*

The crack initiation was defined by the inflection point of the electric potential–COD curve in the electric potential method. As to the measuring system of this study, some electrical noise was observed, causing some uncertainties in deciding the crack initiation, namely determination of the inflection point. These trends were conspicuous at a loading rate of over 100 mm/s. Joyce and Schneider [2] have also reported difficulties in using the d-c potential method in dynamic tests on carbon steel.

With a view of applying the electric potential method to the continuous measurement of crack length in the fracture toughness test at high temperature in the future, a calibration between electric potential and Δa will be prepared using the method that has been reported by Schwalbe and Hellmann [3].

As described above, the present electric potential method still has some problems on the preciseness of deciding the crack initiation point. However, the method is also obviously effective in measuring J_{1d}.

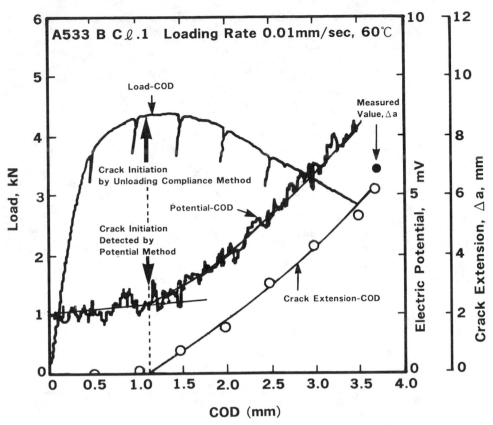

FIG. 9—*Load–COD, electric potential–COD, and crack extension–COD curves.*

Effect of Loading Rate and Temperature on Dynamic Fracture Toughness

The plane-strain fracture toughness K_{IC}, which is the measure of resistance against brittle crack initiation at transition temperature range, depends on loading speed; the transition phenomenon of the fracture toughness is shifted to the higher temperature as loading speed increases, as reported by the authors [4].

However, it is not clear yet how the dynamic fracture toughness J_{Id} is affected by loading rate in terms of the ductile crack initiation in the upper shelf region. Therefore, the authors studied the temperature dependency of the fracture toughness by changing loading rate at three levels for A 533 Gr.B Cl.1 after it was determined that J_{Id} could be effectively measured by the unloading compliance method, as described above. The result is shown in Fig. 11. At this time, the test in the transition temperature range was conducted according to ASTM E 399-83 ANNEX A7. The loading rate (stroke speed) in use was 500 mm/s, 1 mm/ s, and 0.01 mm/s, respectively. K rates measured in the above were, on the average, 2.8 × 10^4 MPa $\sqrt{m}$/s, 8.4 × 10 MPa $\sqrt{m}$/s, and 1.55 MPa $\sqrt{m}$/s. 1.55 MPa $\sqrt{m}$/s conforms to the specified range of ASTM E 399-83 as a static fracture toughness test.

In the case where a valid K_{IC} could not be obtained and stable crack extension became

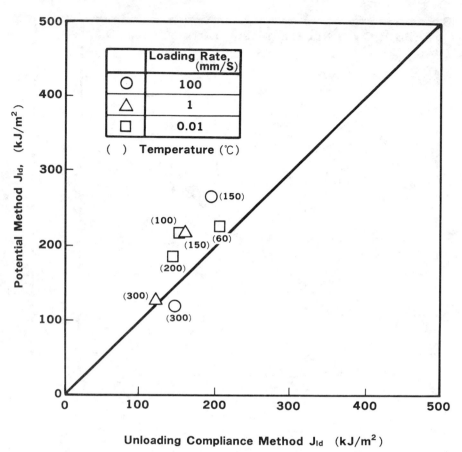

FIG. 10—*Comparison of* J_{Id} *obtained by the potential method with* $J/_{Id}$ *obtained by the unloading compliance method.*

unstable only at an extremely small growth (evaluation of ASTM E 813-81 cannot be applied), the J integral value at instability point was obtained as J_{cd} and then converted to K_{Id} by using Eq 3. In addition, J_{Id} in the upper shelf range was converted to K_{Id} by Eq 4 plotted in Fig. 11. These conversions were applied from the sense that the relation between J_{Id} and K_{Jd} will lie on the continuous extrapolation of the relation between J_{Ic} and K_{Ic} at static condition.

$$K_{Jd} = \sqrt{\frac{J_{cd} \cdot E}{1 - \nu^2}} \tag{3}$$

$$K_{Jd} = \sqrt{\frac{J_{Id} \cdot E}{1 - \nu^2}} \tag{4}$$

where

E = Young's modulus, and
ν = Poisson's ratio.

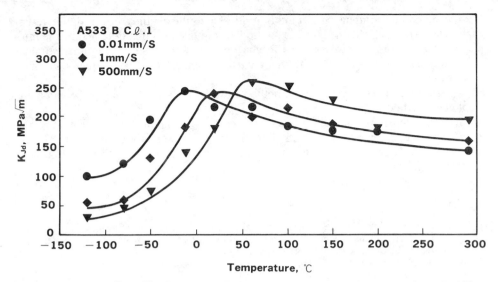

FIG. 11—*Effect of loading rate and temperature on dynamic fracture toughness (A 533 Gr.B Cl.1).*

Referring to Fig. 11, as loading rate increases, the transition curve shifts to the higher temperature side by about 40°C at 1 mm/s and about 70°C at 500 mm/s from a reference transition curve of 0.01 mm/s obtained by a static fracture toughness test. The temperature shift at 500 mm/s calculated by a shift prediction Eq 5 [4], which was already proposed by the authors, is 87°C. Therefore, these shifting quantities might have reasonable correspondence.

$$l_n (\Delta T) = -0.0027\sigma_{ys} + 5.6 \tag{5}$$

where

ΔT = temperature difference between the static fracture toughness transition curve
and the dynamic fracture toughness transition curve, °C, and
σ_{ys} = static yield point at room temperature, MPa.

The fracture toughness in the upper shelf range tends to increase as the loading rate increases, as compared to the static fracture toughness.

In fact, the transition curve of the dynamic fracture toughness is easily obtained for engineering purposes by shifting the static fracture toughness curve by a quantity estimated from Eq 5 to the high temperature.

Conclusion

The effect of the loading rate on the dynamic fracture toughness in regard to the ductile crack initiation is studied by comparing the unloading compliance method with the multiple specimen method using a high-speed servohydraulic testing machine. A method for measuring J_{ld} using the electrical potential method was also examined. Results obtained are as follows.

1. The unloading compliance method is applicable to the ductile crack initiation test at a high loading rate. J_{Id} obtained by this method is equivalent to that obtained by the multiple specimen method. However, the J-R curve obtained by this method underestimates the fracture resistance compared with that measured by the multiple specimen method as Δa increases.

2. J_{Id} based on the electric potential method is about the same as that of the unloading compliance method.

3. J_{Id} varies depending on the loading rate, increasing as the rate increases and remaining conservative compared to the static fracture toughness. The dynamic fracture toughness curve is easily predicted from the static fracture toughness curve both of the transition and upper shelf regions.

References

[1] Saxena, A. and Hudak, S. J., Jr., "Review and Extension of Compliance Information for Common Crack Growth specimens," *International Journal of Fracture,* Vol. 14, No. 5, Alphen an den Rijn, The Netherlands, October 1978, pp. 453–468.

[2] Joyce, J. A. and Schneider, C. S., "Application of Alternating current Potential Difference to Crack Length Measurement During Rapid Loading," NUREG/CR-4699, U.S. Naval Academy, Annapolis, MD, August 1986.

[3] Schwalbe, K. H. and Hellmann, D., "Application of the Electrical Potential Method to Crack Length Measurements Using Johnson's Formula," *Journal of Testing and Evaluation,* Vol. 9, Philadelphia, PA, May 1981.

[4] Sunamoto, D., Satoh, M., and Funada, T., "Study on the Dynamic Fracture Toughness of Steels," *Mitsubishi Technical Review,* Vol. 12, No. 2, Tokyo, Japan, 1975, pp. 71–77.

Narayanaswami Ranganathan,[1] *Gilles Guilbon,*[2]
Khemais Jendoubi,[3] *Andre Nadeau,*[4]
and Jean Petit[5]

Automated Fatigue Crack Growth Monitoring: Comparison of Different Crack-Following Techniques

REFERENCE: Ranganathan, N., Guilbon, G., Jendoubi, K., Nadeau, A., and Petit, J., **"Automated Fatigue Crack Growth Monitoring: Comparison of Different Crack-Following Techniques,"** *Factors That Affect the Precision of Mechanical Tests, ASTM STP 1025,* R. Papirno and H. C. Weiss, Eds., American Society for Testing and Materials, Philadelphia, 1989, pp. 77–92.

ABSTRACT: A software has been developed to conduct constant amplitude fatigue crack growth tests on the lines of a standard test method proposed by the American Society for Testing and Materials (ASTM E 647).

In this paper, after a brief description of this software, the results of calibration tests conducted using compact tension specimens of the high-strength aluminum alloy 2024-T351 are presented and discussed.

The crack length was monitored by the compliance technique and the d-c potential drop technique for automated tests and by optical means for manual verification.

Analysis of the results shows that both crack-following techniques are suitable for automated crack growth monitoring. The maximum absolute error for the compliance technique is on the order of 2.5% and on the order of 1.3% for the potential drop technique.

Calibration functions relating the crack length to normalized compliance for measurements under the loading axis and to the potential drop have been determined for the studied specimen geometry.

Finally, the experimental results relating the crack growth rate to the amplitude of the stress intensity factor by the different techniques used are compared.

KEY WORDS: fatigue crack propagation, crack length measurement techniques, compliance method, potential drop method, calibration functions, crack closure, crack growth rate determination, precision

Fatigue crack propagations tests are conducted to determine the relationship between the crack growth rate da/dN and the cyclic stress intensity factor, ΔK. These tests permit the characterization of the resistance to fatigue crack growth for the material under the loading and environmental conditions studied, and the experimental results can then be used by the designer.

Considering the importance of these tests, ASTM has proposed a standard test method for measurement crack growth rates [1].

---

[1] Maître de Conférences, E.N.S.M.A. Poitiers Cedex 86034, France.
[2] Graduate student, E.N.S.M.A. Poitiers Cedex 86034, France.
[3] Research Chercheur, E.N.S.M.A. Poitiers Cedex 86034, France.
[4] Ingénieur d'Etudes, E.N.S.M.A. Poitiers Cedex 86034, France.
[5] Directeur de Recherche, E.N.S.M.A. Poitiers Cedex, 86034, France.

These tests are carried out either on standard compact tension (CT) or on center cracked (MT) specimens, and in the case of repetitive tests or for tests at low growth rates, manual determination of the da/dN-ΔK relationships can be tedious and time consuming. Moreover, in manual tests, the operator may have to interrupt the tests for various reasons, which can lead to dispersion in the test results, especially in active environments [1,2].

To overcome these difficulties, an automated test method has been developed to carry out such tests conforming to the above-mentioned standard [3,4].

In this paper, the highlights of this software are presented first.

Second, a critical comparison is made between the three different crack-following techniques used in this study and their relative merits discussed.

Finally, the fatigue crack growth resistance curves obtained by the different methods are compared.

Experimental Details

The tests described here were carried out on the aluminum alloy 2024-T351. The composition and nominal mechanical properties of this alloy are given in Tables 1a and 1b, respectively.

Compact tension specimens 75 mm wide and 12 mm thick were used in this study. Figures 1a and 1b show the dimensions of the specimens studied. The specimen shown in Fig. 1b is similar to the standard compact tension specimen [1] except for the dimensions of the initial notch. In one configuration (Fig. 1a) the crack opening displacement, COD or δ, was measured at the load line, and in the second one (Fig. 1b) COD was measured at the front face of the specimen. In these figures the points of current input and potential drop measurement are also indicated. For the potential drop technique, a constant direct current input of 5 A was used. A gain of 1000 was used for the voltage measurements, and the potential drop output was connected to the A-D interface board of the computer.

Two kinds of tests were conducted:

1. Calibration tests at constant load amplitude which were meant to provide the relationship between the crack length and the measured parameters.

In these tests, firstly, the compliance C is defined as equal to δ/P; P being the load was calculated by the computer using an algorithm described later and also determined manually by using an XY plotter. The crack length was optically monitored using a traveling microscope X25.

TABLE 1a—*Nominal composition.*

Element	Si	Fe	Cu	Mn	Mg	Cr	Zn	Ti	Al
% Weight	0.1	0.22	4.46	0.66	1.5	0.01	0.04	0.02	Remaining

TABLE 1b—*Average mechanical properties.*

Yield Strength, MPa	Tensile Strength, MPa	% Elongation (Gage), Length 30 mm	Cyclic Yield Strength, MPa
300	502	11	500

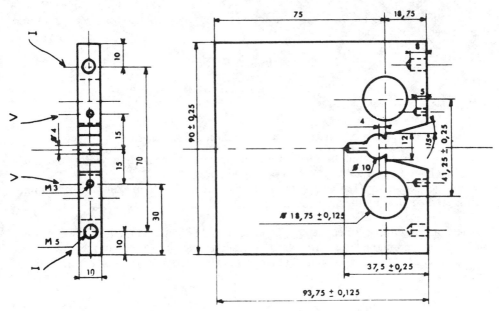

FIG. 1a—*Modified compact tension specimen. Points I and V indicate current input and potential drop measurement points.*

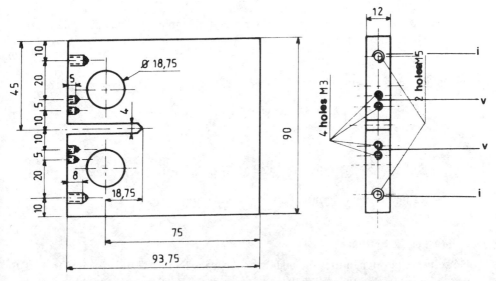

FIG. 1b—*Standard compact tension specimen. Points I and V as in Fig. 1a.*

A function relating the crack length a and the nondimensional compliance *(comp)* was determined. *Comp* is defined as

$$comp = E.B.C.$$ (1)

where

B = the specimen thickness,
E = Young's Modulus = 72 000 MPa for the material used, and
the exact determination of C is presented in detail in the later part of the paper.

Secondly, from potential drop measurements, a calibration function relating the potential drop (V) and the crack length (a) was also determined.

2. In the second series of tests, relationships between da/dN and ΔK were determined for the studied alloy by the different techniques used.

The value of da/dN is calculated by the incremental polynomial method in which a second order polynomial is fitted through $2n + 1$ successive data points. In the present tests n is equal to 2, that is, 5 data points are considered successively. The polynomial is of the form

$$\hat{a} = b_0 + b_1(\zeta) + b_2(\zeta)^2$$ (2a)

where

$\hat{a}$ = the average crack length determined by the polynomial approximation

$$\zeta = (N_i - C_1)/C_2$$

where

N_i = the number of cycles corresponding to a data point a_i, and
$C_1 = 0.5(N_{i-n} + N_{i+n})$, and
$C_2 = 0.5(N_{i+n} - N_{i-n})$.

The crack growth rate, da/dN, is determined by differentiating this polynomial, that is

$$(da/dN)\hat{a} = (b_1/C_2) + 2b_2\zeta/C_2^2$$ (2b)

The value of K is obtained by the following relation [1]

$$K = P/(B\sqrt{W}) \cdot f(\alpha)$$ (3a)

where

$\alpha = a/W$, and
$f(\alpha) = [(2 + \alpha)/(1 - \alpha)^{3/2})](0.886 + 4.64\alpha - 13.32\alpha^2 + 14.72\alpha^3 - 5.6\alpha^4)$ (3b)

This expression for the determination of K is valid for $\alpha > 0.2$ [1].

Description of the Software

The software developed uses the high level language INTRAN incorporated in a PDP-11 computer coupled to the testing machine [6]. It is designed to carry out essentially constant amplitude fatigue crack growth tests under the conditions prescribed by the operator, which are:

1. The maximum and minimum loads during cycling.
2. The test frequency.
3. The number of cycles representing the test duration.
4. The maximum crack length at the end of test.
5. The load maintenance accuracy, which represents the maximum allowable difference between the load attained and demanded.
6. The initial number of cycles between each data acquisition.

The test machine and the analogue input channels of the computer are initially calibrated before starting the test. Once the specimen is fixed in the test grips, the machine control is passed over to the computer, which then prompts the operator to define the above-mentioned test conditions as shown in the example given in Table 2.

The test is then started. During the test the program constantly verifies that the correct loads are maintained, and if at a particular moment the difference between the loads attained and demanded is greater than the demanded precision, a correction is automatically made in the input to the signal generator. Loads are maintained with an accuracy of about 2 bits, which corresponds to about 5 Kg with a load cell capacity of 5000 Kg. This correction becomes frequent at large crack lengths when the system response varies fast as the specimen compliance becomes very high.

The program counts the number of cycles, and the different parameters are measured if the number of cycles corresponds to a data acquisition point. For the second series of tests the data acquisition interval was automatically reduced at the end of tests (at high da/dN values) so that the crack length increment between each acquisition is maintained between 0.2 and 1 mm, as suggested in the standard test method [1].

The test frequency was also reduced at very high da/dN values ($da/dN > 10^{-5}$ m/cycle).

Experimental Results and Analysis

Calibration Tests

As indicated before, these tests are meant to develop relationships between the crack length and the measured parameters, such as the compliance and the potential drop.

Relationship between the compliance and the crack length: Four tests were run to determine the functional relationship between the nondimensional compliance *(comp)* and the crack length *(a)*. In these tests, optical measurements were made on either side of the specimen and checked for differences in crack lengths between the two sides. Differences, if any, were within the limits proposed by the ASTM standards [1].

Manual Determination of Compliance

Compliance was determined manually from δ versus *P* curves, plotted at a frequency of 0.5 Hz on an *X-Y* plotter.

During a fatigue crack propagation test, compliance increases as the crack grows, but the relationship between δ and *P* is not always linear during a loading cycle.

TABLE 2—*Data input to the computer.*

```
***INSTRON*** JOB.06,DPCST6
COD GAGE RANGE (IN MM)=?1,5
FULL SCALE LOAD(IN DAN)=?5000
COD MEASUREMENT ON THE SPECIMEN  EDGE OK = 1, LOAD LINE = 0,0
YOUNGS MODULUS IN MPA = 72000
YIELD STRENGTH IN MPA = 300
SPECIMEN THICKNESS IN MM 10
SPECIMEN WIDTH IN MM 75
DEGREE OF POLY APPROXIMATION FOR P - COD RELATION, MAX = 10,4
THIS PROGRAM WORKS EITHER WITH THE COMPLIANCE METHOD
AND THE POTENTIAL DROP METHOD
INITIAL CRACK LENGTH 18.75
FINAL CRACK LENGTH 60
INITIAL VALUE OF THE POTENTIAL 0.5
CORRECTION FACTOR FOR POTENTIAL1
DO YOU WANT TO RECORD TEST RESULTS, YES = 11
VERIFY THAT THE CONTROL MODE IS CALIBRATED
YES      = 1,      NO = 0


ENTER FILE NAME FOR DATA
*DW1: TEST.1


******THE MACHINE NOW IS IN COMPUTER CONTROL*****
DO YOU WANT TO DO A TEST, YES = 1 1


MAX LOAD (DAN)=300
MIN LOAD (DAN) =3
TEST FREQUENCY (HZ) =30
NB OF CYCLES (MAX=3.2 E+0.7) =1E6
DATA ACQUISITION FREQUENCY N CYCLES 1E4
LOAD MAINTENACE ACCURACY,%F.SCALE LD,(MIN = 0.1% )=0.1
```

Nonlinear effects are introduced due to:

1. Crack closure, induced by plastic deformation in the wake of the crack tip, diminishes the compliance in the initial part of the loading cycle. This effect is predominant at low R ratios [7–10].

2. Plasticity effects at high K_{max} values can lead to an increase of compliance [7].

3. Asperities contact, which can occur because of localized Mode II deformation, introduces a decrease in compliance near the maximum load [8].

To determine the exact value of the compliance corresponding to a particular crack length, it is hence necessary to identify the part of the loading cycle where the δ versus P trace is strictly linear (free from crack closure and the above-mentioned nonlinear effects).

The following method was used for the manual determination of the compliance, which is known as the offset procedure.

In this method, a quantity αP is subtracted electronically from the signal corresponding to the δ to obtain the differential crack opening δ' [10], that is

$$\delta' = \delta - \alpha P \qquad (4)$$

The value of α can be chosen by manipulating a potentiometer.

It can be shown that the slope of the δ' versus P curve reaches zero if α is equal to C, the compliance of the specimen with the crack fully open, for the present tests.

For the test conditions in the present study, that is, for moderate K_{max} values, plasticity effects which can lead to an increase of compliance near the maximum load P_{max} were not observed. Therefore, during a loading cycle and at a given crack length, a decrease in compliance was observed near the minimum load P_{min} due to crack closure and near the maximum load, probably due to asperities contact [8], that is, at either end of the loading cycle, the local compliance C' is smaller than C (defined above). Thus, by differentiating Eq 4, for $\alpha = C$

$$d\delta'/dP = d\delta/dP - C \qquad (4a)$$

The quantity $d\delta/dP$ represents the local compliance C' at a particular point of the loading cycle. Since C' is less than C for the above-mentioned reasons, the slope of the δ'-P curve should be negative at either end of the loading cycle. Also, when the crack is fully open, that is, when the local compliance is equal to C, the slope δ'-P curve should be equal to zero.

In Fig. 2, examples of the δ versus P and δ' versus P diagram for a ΔK of 11 MPa$\sqrt{m}$ at an R value of 0.1 are shown and the crack opening load P_o and the load P' where asperities contact occurs are indicated. The three phases observed in this figure correspond to:

1. Phase I, from the minimum load, P_{min} to P_o where the crack gradually opens.
2. Phase II, from P_o to P', where the crack is fully open.
3. Phase III, from P' to the maximum load, P_{max}, where secondary effects lead to a decrease in compliance.

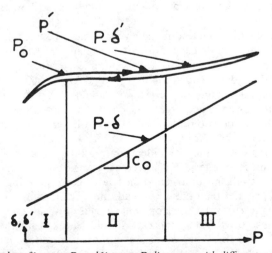

FIG. 2—*Examples of δ versus P and δ' versus P diagrams with different points of inflexion.*

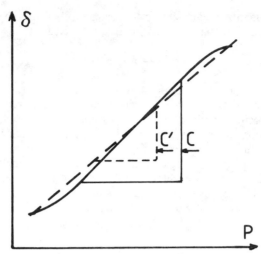

FIG. 3—*Underestimation of the compliance* C *due to closure and other nonlinear effects.*

From these diagrams, it is evident that the exact determination of the crack opening load is a prerequisite for the determination of the true specimen compliance corresponding to the crack length.

If one determines the compliance, by simply taking the difference between the COD values at P_{min} and P_{max} and by dividing it by the load amplitude ΔP, one underestimates the compliance value, as shown in Fig. 3 [11].

For the manual determination, compliance was measured as the slope of the δ versus P curve between points P_o and P'.

Automated Compliance Determination

The signals corresponding to the crack opening δ and the load P were fed into A-D interface board.

At a load acquisition point, the test frequency was reduced automatically to 0.05 Hz (the same as that used for manual determination of compliance). Data couples corresponding to δ and P were read successively by using suitable commands incorporated in the INTRAN language. This data acquisition was limited to the increasing part of the cycle only. The number points for one half cycle were about 150. A polynomial relation was then fitted to pass through the acquired data points by a least square technique. It was generally observed that a polynomial of the fourth order gave a satisfactory fit, a result which has also been observed before, as shown in Fig. 4 [9,12].

The compliance was determined by differentiating this polynomial at the mean load. This definition is valid for the tests considered as it was observed that the crack opening load was always less than the mean load and the load P' corresponding to the second inflexion point was always greater than the mean load under the testing conditions.

The assumption concerning the crack opening level is only valid for constant amplitude tests at medium crack growth rates, and care should be taken for the estimation of this parameter, especially near threshold conditions. In this respect the algorithm presented in Ref 13 is interesting where an iteration technique is used to determine the limits of the linear part of the δ versus P relation. In certain cases, it is also suggested to eliminate data

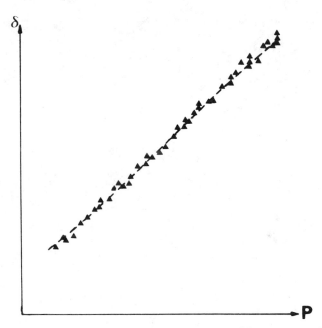

FIG. 4—*Fourth order polynomial approximation of the* P *versus* δ *relation.*

points near P_{min} and P_{max}, where significant scatter is observed due to "rounding off" effects [*9,12*].

Comparison Between the Two Compliance Measurements

Figure 5 compares the value of *comp* estimated by the computer with that determined manually for two tests using the specimen geometry shown in Fig. 1*a*. It can be observed that the computed values of this parameter compare well with the manually determined ones with an average difference of 3.0%.

From these results, the following relation is proposed between *comp* and the crack length *a* for this specimen geometry

$$a/w = 0.1281178 + 0.013775444 \cdot comp - 1.23188 \cdot 10^{-4} \cdot comp^2 \\ + 3.827611 \cdot 10^{-7} \cdot comp^3 \qquad (5)$$

This relationship is valid in the range $0.2 < a/w < 0.7$. In Fig. 6 we have compared the present results with those given in Refs *14* and *15* and find that there is an acceptable correlation between the present results and those in the literature. Thus we consider that the algorithm developed for the determination of the specimen compliance is valid.

For the specimen geometry given in Fig. 1*b*, we used that relationship given in Ref *14* for the determination of the crack length, that is

$$a/w = 1.0010 - 4.6695 \cdot U + 18.460 \cdot U^2 - 236.82 \cdot U^3 + 1214.9 \cdot U^4 - 2143.6 \cdot U^5 \qquad (6)$$

where

$$U = 1/(comp^{0.5} + 1).$$

This expression is valid for $a/w > 0.2$ [*14*].

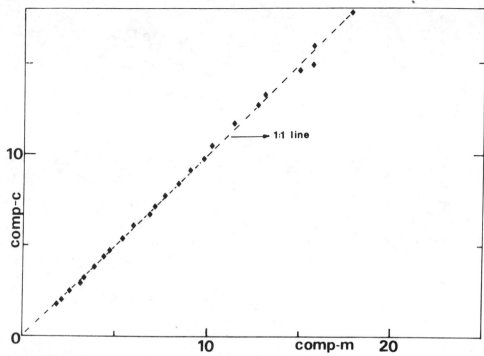

FIG. 5—*Comparison between computed (*comp-c*) and manually estimated (*comp-m*) of the nondimensional compliance.*

Comparison of the crack length estimates between the compliance and the optical techniques was quite good, with an average difference of 1.97 ± 2.65%. The high value of the standard deviation is attributed to the fact that in rare occasions the polynomial approximation for the estimation of the relationship between δ and P failed because of scatter in the acquired data points. After elimination of these data points, the average error estimated is 1.53 + 0.97%.

D-C potential drop technique. This technique was used for subsequent tests using the specimen geometry 1*b*. It should be noted here that the current input points and the potential drop measurement points correspond to the configuration often used for this geometry [*16,17*]. The potential measurements were made at the mean load, thus avoiding crack closure effects.

Four tests were run and measurements of the potential drop, v, made manually using a digital voltmeter. Concurrently, the same signal was fed to the A-D interface board and the corresponding digitized value read by suitable commands. For these computerized measurements, the corresponding input channel was initially calibrated by feeding known values of the potential. The value of the potential drop measured by the computer was within 0.54 ± 0.32% of those measured by the voltmeter, and the following relationship is proposed for the determination of the crack length for 0.3 < a/w < 0.7.

$$a = a_0 + \Delta a \tag{7}$$

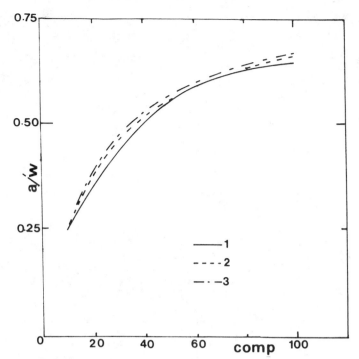

FIG. 6—*Relationship between* comp *and the normalized crack length,* a/w: *curve 1 indicates present results; curve 2 from Ref* 15; *curve 3 from Ref* 14.

where

$$\Delta a = 0.01698 + 37.01579\ \Delta V - 50.87819\ \Delta V^2 + 67.30066\ \Delta V^3 - 40.16394\ \Delta V^4 +$$
$$3.95026\ \Delta V^5 + 4.28112\ \Delta V^6 + 0.52837\ \Delta V^7 - 1.33788\ \Delta V^8 + 0.28286\ \Delta V^9,$$

a_0 = the initial crack length equal to 18.75 mm for the present tests, and

ΔV (in Δa above) = $V - V_0$, where V is the current value of the potential and V_0 is the value of v corresponding to a_0.

This relation is different from the ones proposed in the literature in which it is suggested to use the potential measurements in the nondimensional form, that is, by using the parameter V/V_0 [18]. This modification can be easily brought into the present algorithm.

In Fig. 7 we have compared optical and computed crack lengths by the compliance and potential drop techniques for a test conducted at an R value of 0.33. It can be noted that there is a good correlation between the three measurements. The computed values by the potential drop technique are within 0.78 ± 0.52% as compared to optical measurements.

In Figures 8a, 8b, and 8c we have given the relationship between da/dN and ΔK obtained by the different techniques for a test conducted at an R value of 0.33, and it can be observed that the scatter is slightly more important for the compliance technique especially at low ΔK values.

Based on these observations, it was decided to use the potential drop method for further tests and Fig. 9 gives the da/dN versus ΔK relationship for four values of R for the studied alloy from 14 tests. These results are similar to what is reported in the literature for the studied alloy [19,20].

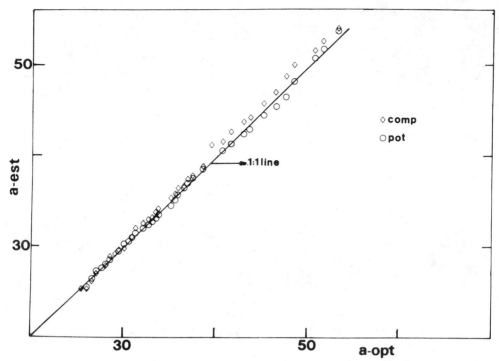

FIG. 7—*Relationship between the estimated value of the crack length* (a-est) *and manually determined ones* (a—opt).

Discussion

In this section we shall discuss the relative merits of the different techniques and sources of error and ways of remedying them.

1. *Optical method:* This is evidently the simplest method and has very little practical limitations except for tests under aggressive environments or at high temperature where the crack cannot be optically monitored. Sources of error are mainly due to crack front curvature, which can lead to an underestimation of the crack length. One way of remedying this is by marking the crack front by slight overloads or by injecting a colored ink to mark the crack front. A correction is then brought into the crack length and eventually the K value if necessary [1].

Differences in crack lengths between the two sides of the specimen can be significant for threshold tests, and care should be taken in the interpretation of the test results.

2. *Compliance method:* In the present results, it has been shown that the estimated values of the crack length by this technique are comparable with the optical measurements within 3%. The advantage of this method is that after an initial calibration it can be utilized in any environment without optically monitoring. Moreover, crack front curvature effects are automatically taken into account as the compliance is related to the average crack length along the crack front.

An inconvenience of this method is that it requires decreasing the test frequency at each point of data acquisition, thus increasing the test duration. Moreover, if environmental effects are predominant, this factor can be a source of error. A possible remedy is the use of dynamic transducers such as strain gages [21].

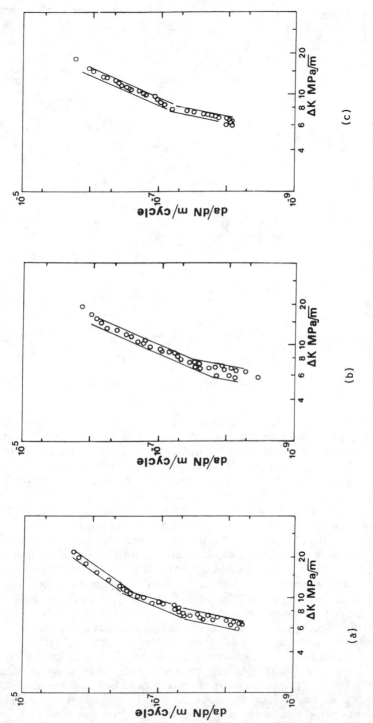

FIG. 8—da/dN versus ΔK curves: (a) optical method; (b) compliance method; (c) potential drop method.

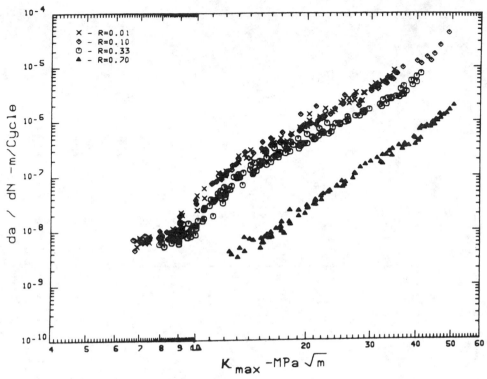

FIG. 9—*Crack growth curves for four R values for the 2024 T-351 alloy.*

Sources of error are related to the exact determination of the compliance, and if crack closure and other effects are not taken into account, it leads to an underestimation of the compliance. Another source of error is due to the discretization of the measurement, and this depends on the system used. Due to the digitized measurements, a scatter is present in the δ versus P diagrams as in Fig. 4, and a polynomial approximation is necessary to determine a smooth curve. The errors in the estimation of this polynomial coupled with those in Eqs 5 and 6 relating the crack length to the compliance can be additive.

In certain computerized tests, it was noticed that estimated values were either systematically greater than or smaller than the optical measurements. This error arises from matching errors in the knife edge supports for mounting the COD gage or due to improper mounting of the gage. In such cases, a correction factor was added to adjust the nondimensional compliance in such a way as to obtain the computed crack length equal to the optically measured one at the beginning of a test. This correction factor was found to remain constant throughout the test.

3. *Potential drop method:* This method is also quite versatile after an initial calibration. In the present tests the calibration tests used optical measurements as the reference values, in which case correction for crack front curvature has to be taken into account. The use of machined slots and electrical analogy methods have also been suggested as calibration techniques [16,17,22]. Once this correction is incorporated, this technique also computes an average crack length. It can be used in most environments except at high temperature where the electrical resistance can be affected. Evidently this method is limited to con-

ducting materials. The use of a conductive coating for nonmetallic materials has been previously suggested [16,17,22].

In the present tests the cycling was stopped to make the potential measurements, and an average value of 15 measurements was made. The tests were much faster than for the compliance method. More time can be saved by making the measurements without stopping the fatigue cycling. One source of error is due to discretization problems, which limit the sensitivity of this technique. For the present tests it is on the order of 1.5 mV.

As in the compliance method, machining errors in the current input and the measurement points are sources of error. This can be remedied by introducing an appropriate correction factor.

Conclusions

In this paper a comparison is made between different crack-following techniques for the development of automated fatigue testing. The following conclusions are drawn from the present results:

1. The compliance method and the potential drop technique can be used for automated crack growth monitoring.
2. Calibration functions relating the nondimensional compliance and the potential drop to the crack length for the specimen geometries used have been determined.
3. The potential drop technique is more precise in the estimation of crack length.
4. The sources of error in automated tests arise mainly from discretizing of data measured and from errors in the specimen machining.
5. These errors can be remedied by introducing appropriate correction factors.

References

[1] ASTM Test Method for Measurement of Fatigue Crack Growth Rates (E 647-86a), American Society for Testing and Materials, Philadelphia.
[2] Wei, R. P. et al., *Metallurgical Transactions,* Vol. 11A, 1980, pp. 151–158.
[3] Guilbon, G., Diplome d'Etude Approfondie (Masters Thesis), E.N.S.M.A., Poitiers, France, 1986.
[4] Ranganathan, N. et al., "Constant Amplitude Fatigue Crack Growth in the Aluminum Alloy 2024 T351," E.N.S.M.A., Poitiers, France, 1986.
[5] Landes, J. D. and Begley, J. A., *Fracture Analysis, ASTM STP 560,* American Society for Testing and Materials, 1974, pp. 170–186.
[6] Instron Software Manual, INSTRON Limited, High Wycombe, Bucks, U.K., 1979.
[7] Elber, W., *Damage Tolerance in Aircraft Structures, ASTM STP 486,* American Society for Testing and Materials, 1974, pp. 230–242.
[8] Ranganathan, N. et al., *Fracture Control of Engineering Structures,* EMAS Pubs., Wareley, U.K., Vol. 3, 1986, pp. 1837–1850.
[9] Ranganathan, N., *Engineering Software for Microcomputers,* Pineridge Press, Swansea, U.K., 1984, pp. 761–762.
[10] Kikukawa, M. et al., *Journal of Material Sciences,* Vol. 26, 1977, pp. 1964–1971.
[11] Sullivan A. M. and Cooker, T. W., *Engineering Fracture Mechanics,* Vol. 9, 1977, pp. 749–750.
[12] Williams, R. S. et al., *Engineering Fracture Mechanics,* Vol. 18, 1983, pp. 953–964.
[13] Sunder, R. *International Journal of Fatigue,* Vol. 7, 1985, pp. 3–12.
[14] Saxena, A. and Hudak, S. J., Jr., *International Journal of Fracture,* Vol. 14, 1978, pp. 456–468.
[15] Newman, J. C., Jr., *Fracture Analysis, ASTM STP 560,* American Society for Testing and Materials, Philadelphia, 1974, pp. 105–121.
[16] Halliday, M. D. and Beevers, C. J., *The Measurement of Crack Length and Shape during Fracture and Fatigue,* Engineering Materials Advisory Services Ltd, Wareley, U.K., 1981, pp. 85–112.

[17] Knott, J. F., *The Measurement of Crack Length and Shape during Fracture and Fatigue,* Engineering Materials Advisory Services Ltd, Wareley, U.K., 1981, pp. 113–135.
[18] Ritchie, R. O. and Bathe, K. J., *International Journal of Fracture,* Vol. 15, 1979, pp. 47–56.
[19] Wanhill, R. J. H., *Engineering Fracture Mechanics,* Vol. 30, 1988, pp. 223–260.
[20] Petit, J. et al., *Proceedings Conf. ICF4, Fracture 1977,* University of Waterloo, Press Canada, 1977, pp. 867–871.
[21] Richards, C. E. and Deans, W. F., *The Measurement of Crack Length and Shape during Fracture and Fatigue,* Engineering Materials Advisory Services Ltd, Wareley, U.K., 1981, pp. 28–68.
[22] Druce, S. G. and Booth, G. S., *The Measurement of Crack Length and Shape during Fracture and Fatigue,* Engineering Materials Advisory Services Ltd, Wareley, U.K., 1981, pp. 136–163.

Charles A. Hautamaki[1]

Resolution Requirements for Automated Single Specimen J_{IC} Testing

REFERENCE: Hautamaki, C. A., **"Resolution Requirements for Automated Single Specimen J_{IC} Testing,"** *Factors That Affect the Precision of Mechanical Tests, ASTM STP 1025*, R. Papirno and H. C. Weiss, Eds., American Society for Testing and Materials, Philadelphia, 1989, pp. 93–102.

ABSTRACT: The results of an investigation are presented which compare the effectiveness of various data acquisition methods for automated single specimen J_{IC} testing. Also discussed are the resolution requirements for obtaining acceptable elastic compliance unload data with 12-bit, 14-bit, and 16-bit analog-to-digital converters (A/Ds) and the use of high-gain window amplifiers.

The results of this investigation indicated that:

1. The system noise has a performance-limiting effect on high-resolution A/D converters.
2. More consistent results were obtained by amplifying the analog signals prior to digitizing than by directly digitizing the signals from high-resolution (16-bit) A/Ds.
3. The number of data pairs used to determine the compliance measured crack length has only a minor effect on the calculated crack length.
4. J_{IC} test results may be adversely affected by using low-resolution data acquisition systems.

KEY WORDS: resolution, J_{IC} testing, compliance, system noise

Resolution requirements of the compliance unload portion of a J_{IC} test have been a concern to those involved in fracture mechanics testing. The assumption has always been that the results would be improved if a higher resolution A/D converter were used. Fundamentally, this is a correct assumption, but as the following discussion will point out, other factors also play an important role in determining what can be considered "good" data.

Procedure

Equipment

The experimental part of this investigation used a 100-kN MTS[2] servohydraulic test system, appropriate control and signal conditioning electronics, 12-bit, 14-bit, and 16-bit analog-to-digital converters, and a DEC[3] PDP 11/23 minicomputer.

---

[1] Senior software engineer, Materials Testing Division, MTS Systems Corp., Minneapolis, MN 55344.
[2] MTS is a registered trademark of MTS Systems Corp., Minneapolis, MN.
[3] DEC is a registered trademark of Digital Equipment Corp., Maynard, MA.

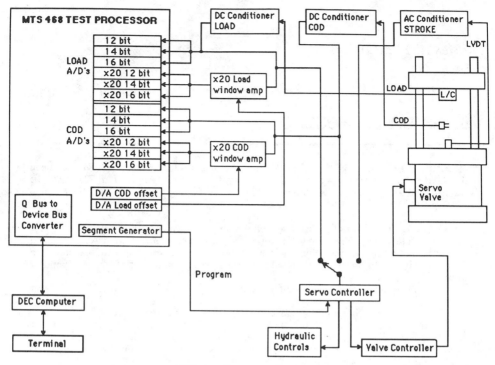

FIG. 1—*System configuration.*

The complete system configuration is shown in Fig. 1. This system is typical of most automated materials testing systems except that three different A/D converters were installed and configured to simultaneously collect the same analog input signals. This was necessary to insure that valid comparisons could be made for data collected. The standard MTS data acquisition system features differential analog inputs, data filtering, zero supression, and simultaneous sample-and-hold circuits, all of which enhance data acquisition and minimize noise problems.

In addition to the standard signal conditioning provided (including a 40-Hz low pass filter), additional window amplifiers were added to increase the gain of the load and COD (crack opening displacement) analog signals prior to digitizing. A three-pole, 25-Hz low pass filter was used on the window amplifier output signal to obtain still lower noise levels. The window amplifiers effectively increased the resolution of the signal by about 4 bits. The amplified signals then became better than 16-bit, 18-bit, and 20-bit resolutions, respectively. The zero offset of the window amplifiers was controlled by a D/A output. The window amplifiers were set to a gain of X20.

Specimen Preparation

The specimens used in this experiment were 20% side-grooved, 1-in., compact-type specimens prepared according to ASTM E 813-81 for J_{IC} testing. The specimens had been precracked to an aspect ratio of approximately 0.625 (a/w) (a = crack length; w = crack width). The material used was HY-180 steel of various heat treats.

Experimental Procedure

The experimental procedure involved three phases. First, static data were collected as a function of data acquisition rates to determine the data variance under static conditions. This procedure consisted simply of collecting load and COD data with each converter and determining the statistical properties of the binary data.

In the second phase, the specimen was repeatedly unloaded to determine its crack length. The load levels used were well below the limit load P_L so that no crack growth would be observed. P_L is defined in ASTM Test Method J_{IC}, A Measure of Fractrue Toughness (E 813-81). The test then ramped in load control to 10 kN, held for 5 s, and unloaded 5% to 9.5 kN at 0.1 kN/s before ramping back to zero load. During the unload, 300 equally spaced, load-COD data pairs were collected to determine the crack length. The transducer data and the window amplifier data were collected simultaneously on all twelve analog-to-digital channels. This hardware configuration allowed the simultaneous collection of load-COD data with an effective resolution from 12 to 20 bits for any single load or COD data point.

Of the data collected during the second phase, the upper 10% and the lower 40% were eliminated, leaving approximately 150 data pairs to determine the slope of the data using a least squares data reduction technique. With the slope determined, the crack length was calculated using the equations described by Saxena and Hudak [1]. The properties evaluated were the crack length, the standard deviation of the crack length, and the correlation of the load-COD slope data [2].

Further analysis of the load-COD data was done to determine what effect the number of data points used in the load-COD slope computation had on the compliance measured crack length. The remaining 150 data pairs were then downsampled and reevaluated using 75, 50, 25, and 12 data pairs.

The third phase of the experiment consisted of running a complete J_{IC} test to determine what overall effect the resolution capabilities would have on a complete test. The tests were run in accordance with ASTM (E 813-81).

Experimental Work

Static Data Collection

This data was used to determine the data variance as a function of data acquisition rates. One thousand points were collected and analyzed at each data rate. The rate was varied from 4 to 100 ms per point. The data was saved in integer format so that over a range of 0 to 65 535 integers, a 12-bit converter could supposedly resolve to within 16 digits, a 14-bit converter to within 4 digits, and a 16-bit converter to within 1 digit. The plots shown in Figs. 2 and 3 indicate the results of 16-bit load and COD data. Table 1 is included as a reference for the equivalent voltage and digit resolution for the analog-to-digital converters.

The noise content for the conditioner output at frequencies less than 100 Hz was generally less than 0.5% of full-scale (25 N and 0.00013 mm, respectively), which is relatively low for most systems. This value can be significantly higher depending upon the surrounding environment and the electrical equipment in use in the facility.

The load data collected during the static test indicated that the statistical variance [3] was generally small, with peaks appearing at multiples of the 60-Hz line frequency. The more erratic COD data were due to signal drift over the test period. These curves more typically exhibited a random variation of values equal to +/- one half the value of the

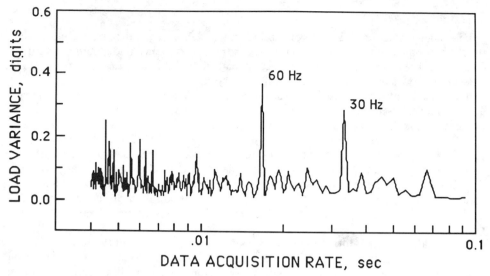

FIG. 2—*Sixteen-bit static load variance as a function of data acquisition rates.*

least significant bit. Thus, the 12-bit converter varied $+/-$ eight digits, the 14-bit converter $+/-$ two digits, and the 16-bit converter $+/-$ one half of a digit.

Compliance Crack Length Measurements

The unload compliance portion of the experiment consisted of: (*a*) performing 200 unloads on a specimen to determine the statistical properties of the compliance measured crack length; and (*b*) determining the correlation of the load-COD curve. This series of tests was performed on three specimens. Specimens A and B were tested using all available A/D channels and the window amplifiers. Specimen C results were obtained using only the standard A/D channels (no window amplifiers). The system noise level at the time of testing varied for unknown reasons. The noise levels were less than 0.3% of full scale before and after the testing of Specimens A and B. The noise level was approximately 0.1% of full scale at the time of testing Specimen C. The same equipment was used for all three specimens, but the tests were conducted several weeks apart. Figure 4 presents the results of the bit resolution versus standard deviation for Specimens A, B, and C. Figure 5 presents the number of points used to determine the compliance slope versus the standard deviation for Specimen A. Table 2 summarizes the numerical values and the correlation coefficient for the tests conducted. In this analysis, the absolute crack length was not relevant since the initial crack length is typically adjusted to the actual modulus of the material, which is often determined from the actual fatigue precracked length.

TABLE 1—*Theoretical resolution available for $+/-$ 10-V data acquisition signals.*

12 bit	4.88 mV	16 digits
14 bit	1.22 mV	4 digits
16 bit	0.31 mV	1 digit

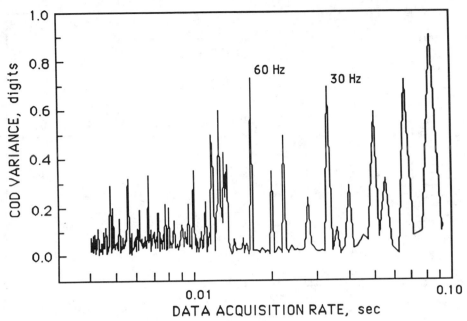

FIG. 3—*Sixteen-bit static COD variance as a function of data acquisition rates.*

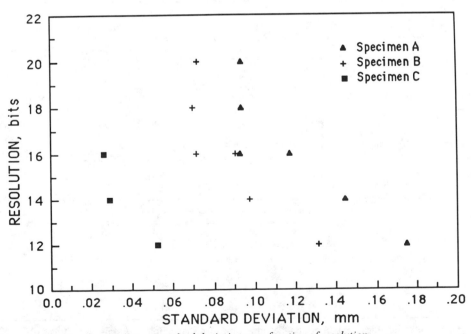

FIG. 4—*Standard deviation as a function of resolution.*

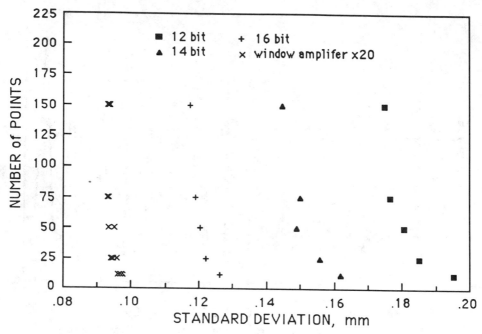

FIG. 5—*Standard deviation as a function of the number of data points.*

TABLE 2—*Average crack lengths and correlations for Specimens A, B, and C based upon 200 observations.*

Specimen	12 Bit	14 Bit	16 Bit	X20 12 Bit	X20 14 Bit	X20 16 Bit
			CRACK LENGTHS, MM			
A	30.686	30.721	30.699	30.588	30.589	30.588
B	30.269	30.227	30.192	30.100	30.100	30.100
C	31.349	31.352	31.355	. . .	. . .	. . .
Specimen	12 Bit	14 Bit	16 Bit	X20 12 Bit	X20 14 Bit	X20 16 Bit
			CORRELATION COEFFICIENTS			
A	0.995772	0.999333	0.999543	0.999919	0.999930	0.999931
B	0.995877	0.999388	0.999592	0.999938	0.999946	0.999949
C	0.999248	0.999838	0.999875	. . .	. . .	. . .

TABLE 3—*Comparison of J_{IC} data using 12, 14, and 16-bit A/Ds.*

Specimen	J_{IC} Values, kJ/m^2			Error Range
	12 Bit	14 Bit	16 Bit	
A	122.1	120.0	119.8	1.8%
B	122.1	120.7	120.4	1.4%
C	89.5[a]	123.3	124.9	28.3%[a]
	123.5			1.3%
D	188.5	188.5	188.2	0.2%

[a] First value includes all data points. Second value eliminates first data point which was obviously in error.

J$_{IC}$ Test Evaluation

This consisted of four complete automated J_{IC} tests with the same data collection techniques used to determine the unload compliance described above. These complete tests did not use the window amplifiers. The results of four tests are shown in Table 3. Example displays of data for Specimen A for 12-bit, 14-bit, and 16-bit evaluations are shown in Figs. 6, 7, and 8, respectively.

Discussion

The intent of this paper is to investigate how significant the use of high resolution A/Ds is in data acquisition for J_{IC} testing. The various phases of this investigation are discussed below.

The investigation began by determining the ability to resolve static data values. From these tests it was determined that the resolution is generally worse than half of the least significant bit accredited to system noise. This is especially true for the lower resolution devices. Rather than obtaining a flat curve as would be expected when the noise is less than the available resolution, the signal varied due to bit toggling (the signal transitioning between available bits). This is apparent in the COD data due to signal drift. The higher bit resolution analog-to-digital converter minimized this effect. This is reflected in the fact that one is able to obtain the harmonics introduced by the line power frequency.

In measuring the unload compliance crack length with the various A/D configurations, a consistent pattern of results appeared. The higher resolution A/Ds gave better results up to a limit. This limit was related to the overall noise in the signal processing system. Figure 4 illustrates that the results continued to improve as the bit resolution increased (as measured by the standard deviation). However, further increases in the resolution beyond the analog noise limit provided no additional capability. Additional capability was obtained

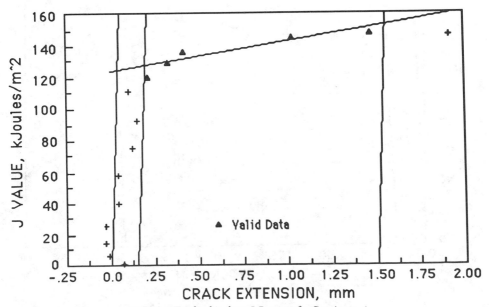

FIG. 6—*Twelve-bit data J-R curve for Specimen A.*

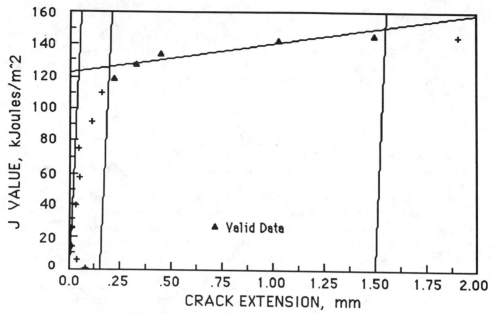

FIG. 7—*Fourteen-bit data J-R curve for Specimen A.*

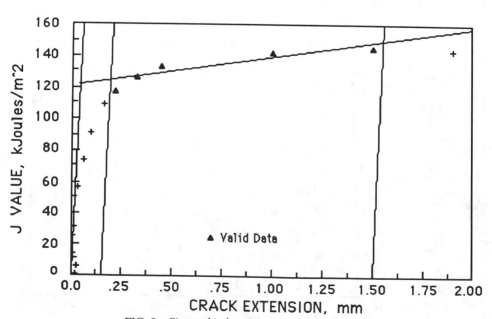

FIG. 8—*Sixteen-bit data J-R curve for Specimen A.*

only when the overall system analog noise level was reduced. The most interesting aspect of the data shown in Fig. 4 is that the 12-bit A/D with the window amplifier offered the same ability to resolve data to this noise limit as did the window amplifier in combination with a 16-bit A/D converter. It should be noted that the window amplifier had a gain of X20 (not X16), and when used with the 12 A/D converter, it gave slightly better results than a 16-bit converter alone.

The number of data pairs used in determining the compliance measured crack length has only a minor effect upon the test results, as shown in Fig. 5. It should be noted though, that there is a greater improvement with more data pairs for the 12-bit data (a 12% change) than there is with the 16-bit data (a 7% change). This figure also illustrates the improvement with higher resolution A/Ds and indicates the noise limit beyond which there is no improvement. Along with the improvement in the standard deviation of the data, the correlation also exhibits the same pattern. Table 2 presents the numerical results of the average crack lengths and correlations for Specimens A, B, and C. The correlation numbers indicate that the data is more consistent when collected using the window amplifiers. There was only a 3.5 to 4% change in the standard deviation for the data collected using the window amplifiers.

The results for the J_{IC} tests indicate that, for the most part, the bit resolution generally does not have a dramatic effect on the final J value. In reviewing the data in Figs. 6, 7, and 8, the data before significant crack extension is more erratic in the 12-bit configuration due to the inability to resolve small changes in crack growth. The consistency of the data improves as the resolution of the A/D increases. Generally, the data between the blunting line and the first offset does not affect the regression analysis unless the initial crack is adjusted. In this analysis, the initial crack length was not adjusted, so the J_{IC} value was not affected. However, in one test case (Specimen C) using a 12-bit converter, the J_{IC} value did change considerably due to a single unfavorable measurement. If this one data point were removed, the data would be consistent for all tests. Because of the possibility of this type of error, it is important to obtain the best data possible for crack length calculations.

Conclusions

From the results of these experiments the following conclusions may be made:

1. The system noise has a limiting effect on high resolution A/D converters. The signal noise level directly affects the available resolution for compliance measurement of crack lengths in J_{IC} testing.

2. Amplifying the analog signals prior to digitizing gives substantially better results than direct digitizing of the signals with high resolution (16-bit) A/Ds. The quality of the data suggests that the resolution of a given analog signal can be more consistently obtained with analog hardware rather than enhanced digital techniques.

3. The number of data pairs of load-COD data used to determine the compliance measured crack length has only a minor effect on the compliance calculated crack length.

4. J_{IC} test values may be adversely affected by using poor data from low resolution data acquisition systems.

Acknowledgments

I would like to thank James L. Maloney, senior research engineer, Latrobe Steel Co., for supplying the specimens used in these experiments. A special note of appreciation is extended to Niel R. Petersen, senior staff engineer, MTS Systems Corp., for his technical assistance in analyzing the results of these experiments.

References

[1] Saxena, A. and Hudak, S. J., *International Journal of Fracture,* Vol. 14, No. 5, Oct 1978, pp. 453–468.
[2] Holman, J. P., *Experimental Methods for Engineers,* 2nd ed., McGraw-Hill, New York, 1971.
[3] Young, H. D., *Statistical Treatment of Experimental Data,* McGraw-Hill, New York, 1962.

Dominique F. Lefebvre,[1] Hassane Ameziane-Hassani,[1] and Kenneth W. Neale[1]

Accuracy of Multiaxial Fatigue Testing with Thin-Walled Tubular Specimens

REFERENCE: Lefebvre, D. F., Ameziane-Hassani, H., and Neale, K. W., **"Accuracy of Multiaxial Fatigue Testing with Thin-Walled Tubular Specimens,"** *Factors That Affect the Precision of Mechanical Tests, ASTM STP 1025,* R. Papirno and H. C. Weiss, Eds., American Society for Testing and Materials, Philadelphia, 1989, pp. 103–114.

ABSTRACT: In this study, we attempt to quantify the effects of certain factors which control the accuracy of multiaxial low-cycle fatigue results obtained from thin-walled tubular specimens. This investigation includes:

1. Some experimental observations regarding the behavior of thin-walled tubes under cyclic axial load and internal-external pressure.
2. A finite-element study of the influence of specimen shape. Tubular specimens of 27-mm diameter with different gauge lengths, wall thicknesses, and transition areas are analyzed. In each case, stress and strain distributions through the wall thickness in the gauge length and the critical strain amplitude at buckling are determined for different combinations of axial load and internal-external pressure.
3. Some experimental fatigue results obtained from tubular specimens under cyclic biaxial loading.

KEY WORDS: low-cycle multiaxial fatigue, tubular specimens, specimen geometry, strain measurement and computation

Nomenclature

D	Inner specimen diameter
L_G	Gauge length
L_T	Transition length
t_o	Wall thickness in the gauge length
t_c	Wall thickness at the specimen extremity
t_T	Wall thickness in the transition area
E	Uniaxial Young's modulus
n	Strain-hardening coefficient
u, v, w, θ	Axial, tangential, radial displacement, and rotation
ε, σ	True strain and stress
ν	Poisson's ratio
e	Engineering strain

[1] Professor, graduate student, and professor, respectively, Department of Civil Engineering, Universite de Sherbrooke, Sherbrooke, Quebec, Canada J1K 2R1.

$( )^e, ( )^p$ Elastic, plastic components
ρ Strain ratio $\rho = \varepsilon_t/\varepsilon_a$
E_a^p, E_t^p Axial, tangential Young's moduli
$( )_a, ( )_t, ( )_r$ Axial, tangential, and radial components

Introduction

Metallic structures and components are generally subject to complex cyclic load histories which eventually may result in fatigue failure. In the study of multiaxial low-cycle fatigue, engineers attempt to derive from simple laboratory test data theories which will permit an adequate assessment of fatigue behavior under complex stress-strain conditions. Many studies have been made to generate valid experimental data and to propose adequate criteria for correlating the available test results.

Three different techniques with different testing systems and specimens are generally used to perform multiaxial fatigue tests: cyclic bending of beams [1], testing of cruciform specimens [2], and tests on thin-walled tubular specimens [3–14]. Among these experimental setups, the most popular employs thin-walled tubular specimens. These have the advantage that axial load, torsion, and internal and external pressure can be applied independently or simultaneously. Furthermore, they allow the possibility of in-phase, out-of-phase, reversible, mean-stress or strain, and hold period tests. Plastic strains are localized in the gauge length, so deformations can easily be measured.

Despite the increasing amount of results published in the literature, accurate correlations with data from different laboratories are often difficult. Many factors may influence the precision and accuracy of results in multiaxial fatigue testing with tubular specimens. They include: material manufacturing and machining, specimen geometry, alignment, electronic controls, strain and stress measurement and computation, definition of failure, data acquisition system, and environment.

Some of the above factors are present in any testing configuration. They are: manufacturing of the material tested, machining and alignment of the specimens, precision, stability and linearity of electronic controls, and reliability of the measuring equipment. They depend on the quality of the material and the testing system, as well as on the care exercised by the experimentalist and the machinist.

This study is limited to the influence of some factors which affect the accuracy and precision of multiaxial low-cycle fatigue results obtained from thin-walled tubular specimens subjected to combinations of axial load and internal-external pressure. We consider, in particular, factors which can be controlled by the researcher, namely the specimen geometry, the methods of strain or stress measurement and computation, and the definition of failure.

Development of Thin-Walled Tubular Specimens

Experimental Observations

Among the various possible specimen designs, the thin-walled tubular specimen has been selected by many workers involved in low-cycle multiaxial fatigue because it enables a fairly uniform stress-strain distribution in the gauge length, even in the plastic range, and because of its facility in measuring and determining stresses and strains. However, the principal disadvantage of thin-walled tubes is their tendency to buckle under relatively low compressive strains.

The final selection of specimen dimensions results in a compromise between the conflicting requirements of uniform strain field at low- and high-strain levels and instability. Geometrical factors which affect the behavior of the specimen are thickness, gauge length, transition area, and fillet radius. Thicker-walled sections increase the resistance to buckling in compression and minimize errors caused by possible eccentricity of axial loading. However, as the thickness is increased, the specimen can no longer be considered as a thin-walled tube, and the derivation of the stresses from applied loads becomes more complex. The resistance to buckling is also influenced by the ratio of gauge length to diameter. Transition area and fillet radii influence the stress and strain distributions as well as crack formation. A specimen which is heavily reinforced outside the guage length will develop a "barrel shape" so that complex bending strains will be superimposed on the system.

Figure 1 shows five different profiles of thin-walled tubular specimens used in low-cycle multiaxial fatigue under combinations of axial load and internal-external pressure. Profile A was designed by Havard [3] for the testing of 1018 mild steel with stress ratios $\sigma_a/\sigma_t = 1.21, 0.50,$ and -0.34. The maximum tangential strain amplitude before buckling is 0.4% for a stress ratio of 1.21 and 0.8% for a stress ratio of -0.34.

Profile B was used by Ellison and Andrews [4] for the testing of an aluminum alloy RR58. The original gauge length and wall thickness were chosen as 50 mm and 1 mm, respectively, but these dimensions gave rise to buckling under external pressure before yield. The final dimensions were determined so that strains up to -0.5% were attainable without buckling; a wall thickness of 1 mm with a gauge length of 22 mm was used for strain ratios of $\varepsilon_t/\varepsilon_a = -1, -0.5,$ and 0.0. The dimensions were changed to 1.25 mm and 19 mm, respectively, for strain ratios of 0.5 and 1.0. Further finite-element analyses conducted by Lohr and Ellison [5] showed the presence of strain peaks at the fillet radius and at the end of the gauge length for combinations of axial load and internal pressure. They used profile C for the testing of 1 Cr-Mo-V steel for strain amplitudes ($\Delta\varepsilon = {}^{+6}_{-0}$) up to 1%. For strain ratios $\varepsilon_t/\varepsilon_a = 1.0$ and -1.0, no premature failure conditions due to buckling or strain concentrations were observed.

Profile D was designed by Found et al. [6] for the testing of an aluminum alloy and steel under combinations of axial load, internal-external pressure, and torsion. A gauge length greater than the inner diameter is required to allow shear strains to develop uniformly inside the gauge length. To avoid buckling, a ratio of inner diameter to outer diameter of 0.80 was chosen. Thus, this specimen may not be considered as a thin-walled tube.

Profile E was used by the present authors for the testing of aluminum and steel in low-cycle multiaxial fatigue (axial force and internal-external pressure). It resulted from a long development process [7,8] and differs from the other profiles in that it has a tapered transition length. The larger gauge length (10 mm) is used for low-strain amplitudes ($\varepsilon_a \leq \pm 0.2\%$), and the shorter one is used for strain amplitudes between ± 0.2 and $\pm 0.5\%$. The limitations and validity of this specimen are evaluated in this study through a finite-element analysis and experimentation.

Finite-Element Study

A finite-element study of specimens with the dimensions shown in Fig. 2, but with varying gauge length, transition length, and wall thickness, has been undertaken. In this study, the Donnell-Mushtari-Vlasov shell theory is employed [15]. The equivalent stresses and strains are computed using J_2 flow theory of plasticity. Because of symmetry, one half of the specimen is analyzed, with a variable thickness (t_o at the center of the gauge length and t_c at the specimen extremity). The half specimen is divided into 26 axisymmetric elements

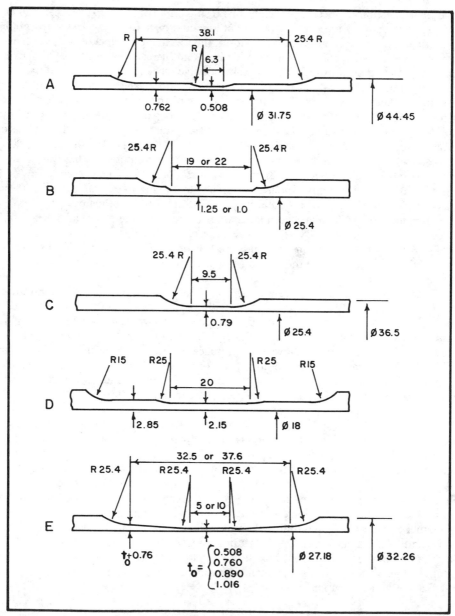

FIG. 1—*Profiles of thin-walled tubular specimens: (A) Havard* [3]; *(B) Andrews and Ellison* [4]; *(C) Lohr and Ellison* [5]; *(D) Found et al.* [6]; *(E) Lefebvre et al.* [8].

each having three nodes with three degrees of freedom per node (displacements u, w, and the rotation θ). Each element has six Gauss integration points along its length and five Gauss integration points across its thickness. Throughout the deformation process, the applied pressure is maintained perpendicular to the deformed wall of the specimen.

The imposed displacements are $w = \theta = 0$, $u \neq 0$ at the extremity and $u = \theta = 0$,

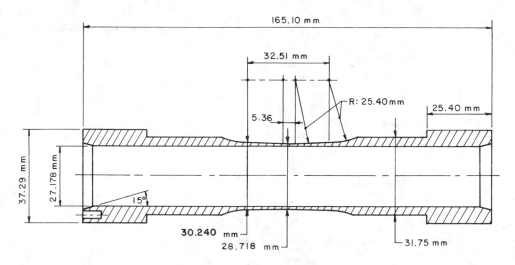

TUBULAR   SPECIMEN

FIG. 2—*Tubular specimen geometry for testing of A516 Steel* [13].

$w \neq 0$ at the center of the gauge length. The axial load eccentricity is approximated by: $e = e_o \cos\pi \ x/L$ with $e_o = (t_c - t_o)/4$, where L is the half length of the specimen. By an incremental process, the finite-element analysis allows us to determine the stress and strain distributions through the thickness and along the length of the specimen, and the axial load and pressure at buckling for different imposed stress ratios and specimen configurations. The different profile dimensions of the specimens considered in the numerical analysis are given in Table 1.

The uniaxial stress-strain law is taken as follows [*15*]

$$\varepsilon = \begin{cases} \sigma/E & \text{if} \quad \sigma \leq \sigma_Y \\ \dfrac{\sigma_Y}{E}\left[\dfrac{1}{n}\left(\dfrac{\sigma}{\sigma_Y}\right)n - \dfrac{1}{n} + 1\right] & \text{if} \quad \sigma \geq \sigma_Y \end{cases}$$

For the A516 steel used in this study, we took $n = 7.5$, $\sigma_Y = 490$ MPa, and $E = 205\ 000$ MPa.

The imposed stress ratios at the beginning of the incremental process were taken as

TABLE 1—*Profile dimensions used in the finite-element analysis.*

Specimen Number	Gauge Length, L_G (mm)	Transition Length, L_T (mm)	Thickness in the Gauge Length: t_o (mm)
1	5.0	13.7	0.508
2	5.0	13.7	1.016
3	10.0	13.7	0.508
4	10.0	13.7	1.016
5	5.0	0.0	1.016

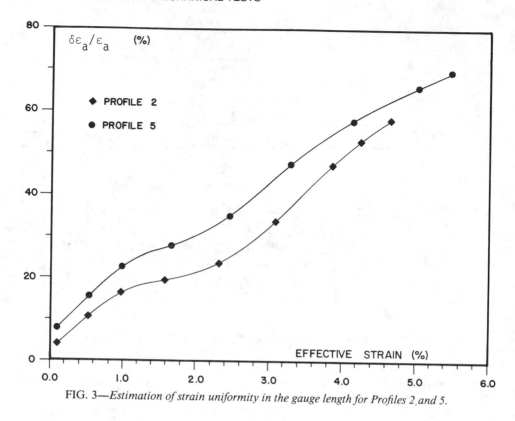

FIG. 3—*Estimation of strain uniformity in the gauge length for Profiles 2 and 5.*

σ_t/σ_a = 0.0, −0.30, and −0.5. Figures 3 to 6 give the results obtained for the stress ratio of −0.5. They show the degree of nonuniformity of stresses or strains inside the gauge length through the wall thickness when the applied effective strain varies from zero to its critical value at buckling. Each value reported was computed at the first Gauss integration point of the first element situated in the gauge length. Figures 3 and 4 show the comparison between the profiles with and without a transition length, the other dimensions being the same. It can be seen that a transition length produces the best uniformity in the stress and strain distributions, but reduces the critical effective strain at buckling. Experimental results show that buckling appears at relatively small strain amplitudes during cyclic loading. For the thin-walled tubular specimen tested, the strain amplitudes are approximately limited to 2% in cyclic uniaxial cyclic loading and to 0.8% in cyclic shear loading (σ_a/σ_t = −1.0).

A comparison of Specimen Configurations 1, 2, 3, and 4 is given in Figs. 5 and 6. Since the resistance to buckling is proportional to the wall thickness and increases when the gauge length decreases, the greater equivalent strain amplitude at buckling is obtained for Specimen 2 (L_G = 5 mm, t_o = 1.016 mm). The most uniform stress and strain distributions through the wall thickness are obtained with Specimens 3 and 4, with a maximum error of 3% when the effective strain amplitude is smaller than 1%. Specimen 4 (L_G = 10 mm, t_o = 1.016 mm) represents the best compromise between resistance to buckling and uniformity of stresses and strains.

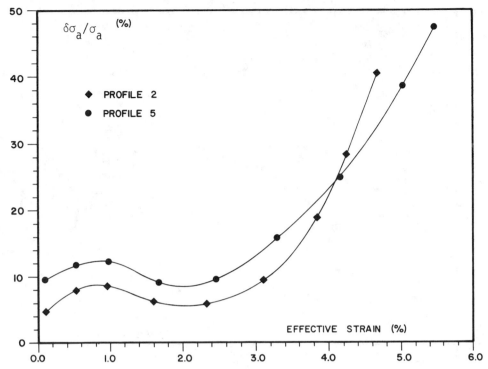

FIG. 4—*Estimation of stress uniformity in the gauge length for Profiles 2 and 5.*

Choice of Specimen Geometry

The choice of specimen geometry is not unique but is affected by testing facilities and other possible limitations. In multiaxial low-cycle fatigue for combinations of axial loading and internal-external pressure, with controlled strain amplitudes limited to ±0.5% for strain ratios of +1.0 and −1.0, practical considerations and the results of the finite-element analysis lead to certain requirements in the definition of the geometry of thin-walled tubular specimens in aluminum alloy or steel:

1. Ratios of inner diameter to outer diameter larger than 0.9 may be specified to simulate thin-walled conditions.
2. Large fillet radii avoid stress and strain concentration near the fillet, but, if these radii are too large, strain peaks may appear at the middle of the gauge length. Radii equal to the inner diameter is a compromise between these two requirements.
3. As shown by the finite-element analysis, a transition length is beneficial for the uniformity of stress or strain. A tapered transition length larger than half of the inner diameter is recommended, the thickness at the end of the transition length being situated between $1.5t_o$ and $2t_o$.
4. Resistance to buckling and uniformity of stress and strain distribution are essentially controlled by the ratios of gauge length to inner diameter (L_G/D) and of gauge length to wall thickness (L_G/t_o). Resistance to buckling requires a small value of L_G/D and a large

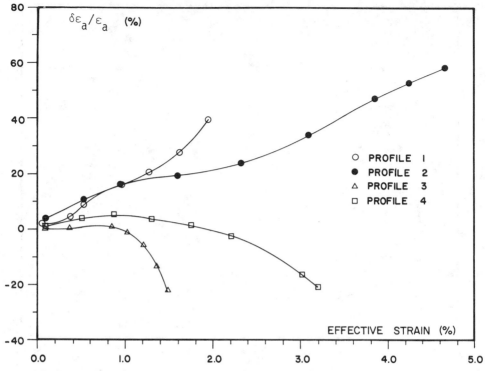

FIG. 5—*Estimation of strain uniformity in the gauge length for Profiles 1, 2, 3, and 4.*

value of L_G/t_o. Requirements of stress and strain uniformity give opposite trends. The finite-element analysis shows that a ratio $L_G/D \geq 0.5$ and a ratio L_G/t_o between 10 and 15 should be a good compromise between these two conflicting requirements.

5. Eccentricity of the applied axial load depends on the ratio t_c/t_o where t_c is the wall thickness at the point of application of the axial load. The finite-element study of Profiles 3 ($t_o = 0.508$ mm, $t_c = 2.54$ mm, $L_G = 10$ mm) and 4 ($t_o = 1.016$ mm, $t_c = 2.54$ mm, and $L_G = 10$ mm) shows this ratio has little influence on stress and strain uniformity since a smaller wall thickness t_o produces the best uniformity inside the gauge length. Practical considerations should limit the ratio t_c/t_o to a maximum value of 4.

6. For the specimen configuration shown in Fig. 2, the above recommendations should lead to the following dimensions: inner diameter $D = 27.18$ mm, gauge length $L_G = 13.6$ mm, wall thickness 0.90 mm $\leq t_o \leq 1.36$ mm, transition length $L_T = 13.6$ mm, radii at gauge length and transition length $R = 25.4$ mm, wall thickness at the end of transition length $t_T = t_o + 0.76$ mm. By extrapolation of the finite-element results on A516 steel, the maximum error in nonuniformity should be 4%, and the critical equivalent strain at buckling should be approximately 3% with $t_o = 1.36$ mm and 2% with $t_o = 0.90$ mm.

Measurement and Computation of Parameters

As multiaxial fatigue testing is generally very complex and expensive, data from different workers published in the literature are frequently used to verify the various proposed the-

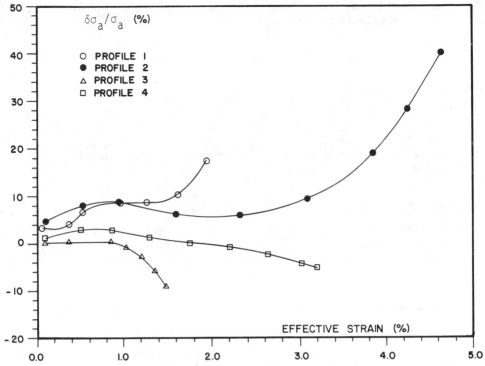

FIG. 6—*Estimation of stress uniformity in the gauge length for Profiles 1, 2, 3, and 4.*

ories. To make this possible, similar testing conditions and hypotheses concerning measurements and the computation of parameters are required.

Fracture is determined when an engineering crack occurs and is defined to be the state where the stress amplitude in any direction becomes unstable. A 5% reduction of stress amplitude is generally adopted by workers using thin-walled tubular specimens.

Strains are ordinarily measured by strain gauges or transducers. Experience has shown that no significant difference is observed when a measurement system is used in air or in oil. When transducers are used, it is recommended that small pieces of synthetic textile material be bonded with adhesive to the specimen at the contact points to reduce the tendency for the transducer to move during assembly and avoid surface damage.

In multiaxial low-cycle fatigue, the conventional strains e_a and e_t are ordinarily controlled and held constant during each test. The derived strain parameters are given by Ref 8

$$\varepsilon_a = \ln(1 + e_a) \qquad \varepsilon_t = \ln(1 + e_t)$$
$$\varepsilon_a^e = \Delta\sigma_a/2E_a^p \qquad \varepsilon_t^e = \Delta\sigma_t/2E_t^p$$
$$\varepsilon_a^p = \varepsilon_a - \varepsilon_a^e \qquad \varepsilon_t^p = \varepsilon_t - \varepsilon_t^e$$

where

$$E_a^p = \frac{E(1 + \nu^e\rho)}{1 - (\nu^e)^2} \qquad E_t^p = \frac{E(1 + \nu^e/\rho)}{1 - (\nu^e)^2}$$

are the axial and circumferential Young's moduli, respectively. The radial strain is not measured, but computed from the relation

$$\varepsilon_r = -\frac{\nu^e}{1 - \nu^e}(\varepsilon_a^e + \varepsilon_t^e) - \frac{\nu^p}{1 - \nu^p}(\varepsilon_a^p + \varepsilon_t^p)$$

In most studies published in the literature, the elastic and plastic strain components are not available and the radial strain is obtained from the incompressibility condition ($\varepsilon_a + \varepsilon_t + \varepsilon_r = 0$). However, this hypothesis, which assumes fully plastic behavior, is not compatible with the cyclic behavior in low-cycle fatigue, where Poisson's ratio varies between its elastic value in the elastic regions of the hysteresis loop and its plastic value in the plastic regions.

Experimental Verifications

A study of multiaxial low-cycle fatigue of A516 Gr 70 steel has been undertaken by the authors. A complete description of the testing device is given in Ref 8, and uniaxial and biaxial fatigue data are reported in Refs 13 and 14. The geometry of the thin-walled tubular

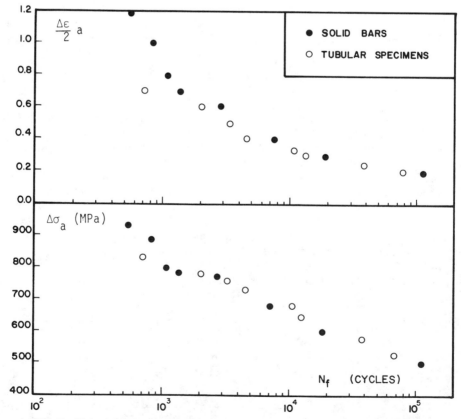

FIG. 7—*Uniaxial fatigue results for A516 steel with solid bars and tubular specimens.*

specimen tested under combinations of axial load and internal-external pressure is represented in Fig. 2. Uniaxial fatigue data obtained from solid bars and thin-walled tubes are given in Fig. 7. It can be seen that, in terms of stresses, the fatigue lives of each type of specimen at a given stress amplitude are approximatively identical, while, in terms of strains, the fatigue lives of tubular specimens are approximatively 10 to 20% less than those obtained from solid bars for a same strain amplitude. In Fig. 8, hysteresis loops at different strain amplitudes for a strain ratio $\varepsilon_t/\varepsilon_a = -1.0$ are represented. Premature failure by buckling appears in the tangential direction for strain amplitude equal to 0.8%.

Conclusion

In this study, we have described some effects of specimen geometry and of techniques for the measurement and computation of strains on the accuracy and precision of multiaxial low-cycle fatigue results obtained from thin-walled tubular specimens under cyclic axial load and internal-external pressure. Experimental observations and a finite-element analysis have led us to formulate some recommendations regarding the dimensions of gauge length, wall thickness, and transition area of a thin-walled tubular specimen. It has also been shown that, when axial and tangential strains are controlled and measured by strain gauges or transducers, an adequate computation of the radial strain is required to correlate the multiaxial low-cycle fatigue results and verify the various fatigue theories proposed in the literature.

Acknowledgments

This work was supported in part by the Natural Sciences and Engineering Research Council of Canada (NSERC) and the Government of the Province of Québec (Programme FCAR).

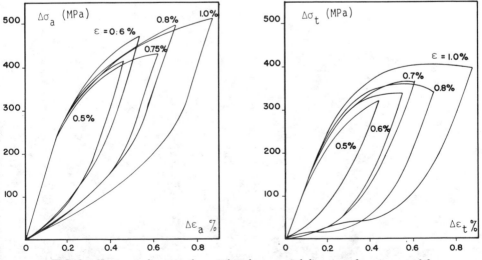

FIG. 8—*Hysteresis loops in the axial and tangential directions for $\varepsilon_t/\varepsilon_a = -1.0$.*

References

[1] Krempl, E., "The Influence of State of Stress on Low-Cycle Fatigue of Structural Materials: A Literature Survey and Interpretive Report," in *The Influences of State of Stress on Low-Cycle Fatigue on Structural Materials: A Literature Survey and Interpretive Report, ASTM STP 549,* 1974.

[2] Pascoe, K. J. and De Villiers, J. W. R., "Low-Cycle Fatigue of Steels Under Biaxial Straining," *Journal of Strain Analysis,* Vol. 2, No. 2, 1967, pp. 117–126.

[3] Havard, D. G., "Fatigue and Deformation of Normalized Mild Steel Subjected to Biaxial Cyclic Straining," Ph.D. thesis, University of Waterloo, Waterloo, Ontario, Canada, 1970.

[4] Andrews, J. M. H. and Ellison, E. G., "A Testing Rig for Cycling at High Biaxial Strains," *Journal of Strain Analysis,* Vol. 8, No. 3, 1973, pp. 168–175.

[5] Lohr, R. D. and Ellison, E. G., "Biaxial High-Strain Fatigue Testing of 1 Cr-Mo-V Steel," *Fatigue of Engineering Materials and Structures,* Vol. 3, 1980, pp. 19–37.

[6] Found, M. S., Fernando, U. S. and Miller, K. J., "Requirements of a New Multiaxial Fatigue Testing Facility," in *Multiaxial Fatigue, ASTM STP 853,* 1985, pp. 11–23.

[7] Ellyin, F. and Valaire, B., "High-Strain Biaxial Fatigue Test Facility," *Proceedings,* the 1982 Joint Conference on Experimental Mechanics, SESA-JSME, Hawaii, 23–28 May 1982, Society of Experimental Stress Analysis, Brookfield Center, CT, pp. 136–143.

[8] Lefebvre, D., Chebl, C., Thibodeau, L., and Khazzari, E., "A High-Strain Biaxial Testing Rig for Thin-Walled Tubes Under Axial Load and Pressure," *Experimental Mechanics,* Vol. 23, 1983, pp. 384–391.

[9] Socie, C. F., Waill, A. and Dittmer, D. F., "Biaxial Fatigue of Inconel 718 Including Mean Stress Effects," in *Multiaxial Fatigue, ASTM STP 853,* 1985, pp. 463–481.

[10] Ohashi, Y., Tanaka, E. and Ooka, M., "Plastic Deformation Behavior of Type 316 Stainless Steel Subject to Out-of-Phase Strain Cycles," *Journal of Engineering Materials and Technology,* Vol. 107, 1985, pp. 286–292.

[11] McDiarmid, D. L., "Fatigue Under Out-of-Phase Biaxial Stresses of Different Frequencies," in *Multiaxial Fatigue, ASTM STP 853,* 1985, pp. 606–621.

[12] Zamrik, Y. S. and Frishmuth, R. E., "The Effects of Out-of-Phase Biaxial Strain Cycling on Low-Cycle Fatigue," *Experimental Mechanics,* Vol. 13, 1973, pp. 204–208.

[13] Lefebvre, D. F., "Hydrostatic Pressure Effect on the Life Prediction in Biaxial Low-Cycle Fatigue," Second International Conference on Biaxial/Multiaxial Fatigue, Sheffield, U.K., 16–20 Dec. 1985, in preparation.

[14] Lefebvre, D. F. and Ellyin, F., "Cyclic Response and Inelastic Strain Energy in Low-Cycle Fatigue," *International Journal of Fatigue,* Vol. 6, 1984, pp. 9–15.

[15] Hutchinson, J. W., "Plastic Buckling" in *Advances in Applied Mechanics,* Vol. 14, 1974, pp. 67–144.

Alignment Problems

Rob J. H. Baten,[1] Hans J. d'Haen,[2] Fred A. Jacobs,[3]
Martin K. Muller,[4] Pieter E. van Riesen,[5]
and Gin Lay Tjoa[6]

Requirements for the Permitted Size of the Alignment Errors of Load Frames for Fatigue Testing and a Proposal for a Relevant Measuring Method

REFERENCE: Baten, R. J. H., d'Haen, H. J., Jacobs, F. A., Muller, M. K., van Riesen, P. E., and Tjoa, G. L., **"Requirements for the Permitted Size of the Alignment Errors of Load Frames for Fatigue Testing and a Proposal for a Relevant Measuring Method,"** *Factors That Affect the Precision of Mechanical Tests, ASTM STP 1025,* R. Papirno and H. C. Weiss, Eds., American Society for Testing and Materials, Philadelphia, 1989, pp. 117–135.

ABSTRACT: This paper covers the activities of the working party on Requirements for Fatigue Testing Machines [*3*].

Requirements and recommendations are specified with respect to the alignment of load frames for fatigue testing, including load cell and (hydraulic) actuator.

Subsequently, a method is proposed to measure the alignment errors. Calculations are made and a series of alignment specimens are designed for load frames of different capacities to check whether or not the bending stresses under load, due to misalignment, correlate with the off-load measured alignment errors. Designs for clamping the alignment specimens are given and applied for the measurements as well.

The alignment errors to be specified, defined, and measured include:

1. Eccentricity.
2. Angular deflection.
3. Alignment error.

Although a literature survey was carried out, there appeared to be no specific requirements for the alignment accuracy of load frames. International standards mainly set out requirements concerning the accuracy of the load cell and the maximum allowable specimen bending at zero load. These standards, however, have omitted the effect of the alignment errors as defined in this paper.

Results of measurements obtained with the described method lead to a recommendation for its application, either in part or in total.

KEY WORDS: test equipment, requirements, fatigue, measurement, error analysis, alignment, samples, positioning

[1] Joint Laboratories and Consulting Services for the Dutch Electricity Supply Companies, NV KEMA, Arnhem, The Netherlands.
[2] Aircraft Factory Fokker, Schiphol-Oost, The Netherlands.
[3] National Aerospace Laboratory, NLR, Emmeloord, presently Royal Huisman Shipyard—Vollenhove, The Netherlands.
[4] Metal Research Institute TNO, Apeldoorn, The Netherlands.
[5] University Twente, UT, Enschede, The Netherlands.
[6] Secretary, Netherlands Energy Research Foundation, ECN, Petten, The Netherlands.

1. Introduction

This paper deals with the requirements for the permitted size of the alignment errors of load frames for fatigue testing and a proposal for a relevant method to measure these errors. The proposed requirements concern the alignment of the load frame, including load cell and (hydraulic) actuator.

To measure the alignment of the frame under load according to this method, an alignment specimen provided with strain gauges was prepared.

The alignment errors with respect to the load frame to be defined and measured include:

1. Eccentricity F (mm).
2. Angular deflection α (rad).
3. Alignment error f_1 (mm).

Although a literature survey was carried out, there appeared to be no specific requirements for the alignment accuracy. International standards mainly set out requirements concerning the accuracy of parts of a testing system like load cells and extensometers and the maximum allowable bending in a specimen at zero load. These institutes, however, have omitted the inaccuracy of the frame that leads to the alignment error f_1 and angular deflection α on the eccentricity F as defined in this paper.

It was deduced that the maximum permissible strain that the specimen undergoes from bending can vary between 3 and 10% of that caused by the axial load, depending on the kind of test. A good survey is also given by J. Bressers [1] and B. W. Christ and S. R. Swanson [2].

To obtain feasible requirements, a number of manufacturers of fatigue testing machines have been contacted to specify the accuracy of the load frames they offer with respect to eccentricity and angular deflection. It was found that their figures were specified mainly at $0 < L \leq 300$ mm and at $L > 300$ mm. A large scatter was found in these specifications, all which relate to standard fatigue testing machines.

It is unknown, however, in which way eccentricity and angular deflection are defined by the manufacturers and which method has been applied to measure these parameters. Therefore, it was necessary to lay down general definitions relating to these specifications.

With respect to the piston rod displacement of the actuator, corresponding specifications appeared to be desirable as well.

This paper is an abridged version of a more detailed report on the same subject, issued by the working party in 1986 [3]. A copy of that report is available on request.

2. Method to Determine the Alignment Error (f_1) and Angular Deflection (α) of the Unloaded Frame to Calculate the Eccentricity (F)

The measuring method as described in this chapter has been carried out with three different types of instruments:

1. Measurements with dial gauges $L \leq 300$ mm.
2. Measurements with electronic clinometers $L \geq 300$ mm.
3. Measurements with an alignment telescope $L \geq 300$ mm.

The combination of their results will give full information on the alignment parameters of the unloaded frame. The total alignment analysis of the frame also includes measurements of the errors of the piston rod displacement.

However, these measurements will be preceded by a general inspection of the load frame. The general inspection will always be executed in combination with measurements with dial gauges as part of the acceptance testing or as part of the quality control scheme.

2.1. General Inspection

It is recommended to submit the load frame to a general inspection before the alignment is measured. This inspection concerns, for example:

1. The finish of the threaded spindles and/or smooth columns and piston rod.
2. Geometry of the load frame, including the position of the load cell and of the actuator relative to the spindles/columns.
3. The functioning of the (hydraulic) clamping device.
4. The finish of the planes used for measuring the alignment, especially with respect to out-of-roundness and flatness of the mounting planes.

2.2. Definitions Necessary to Determine the Eccentricity F

Before describing the measuring methods, the working party defined a number of terms relating to the alignment errors. These include:

1. *Mounting plane* (Fig. 1a). Circular end plane of the load cell and of the piston rod of the actuator to which a clamping device can be attached.
2. *Angular deflection α (radians)* (Fig. 1a). The angle between the mounting planes while the piston rod is in the midstroke position.
3. *Length L (mm)* (Fig. 1b). The distance between the centers of the mounting planes while the piston rod is in the midstroke position.
4. *Eccentricity of the load frame F (mm)* (Fig. 1c). The distance between the intersections of the center normals of the upper and lower mounting plane with the bisector plane bisecting the angle α between the mounting planes while the piston rod is in the midstroke position.
5. *Alignment error f_1 (mm)* (Fig. 1c). The distance between the center normal of the lower mounting plane and the center of the upper mounting plane while the piston rod is in the midstroke position.

2.3. Measurement with a Dial Gauge to Determine the Eccentricity F at L

According to the above-mentioned definitions, the alignment error f_1 and the angular deflection α have to be measured (Fig. 1c). Because α and η are assumed to be very small, η can be a substitute for α in: $\sin\alpha \approx \tan\alpha \approx \alpha$ and $\cos\alpha \approx 1$.

It follows:

$$L_1 = L \cos\eta = L$$
$$a = \tfrac{1}{2}L_1 \tan\alpha = \tfrac{1}{2}L_1 \alpha$$
$$f_1 = F \cos\tfrac{1}{2}\alpha + a \cos\tfrac{1}{2}\alpha = F + a.$$

From substitutions follows: $F = f_1 - \tfrac{1}{2}L\alpha$.

For a direct measurement at $L \leq 300$ mm, it is recommended for practical reasons to measure f_1 and α with a dial gauge.

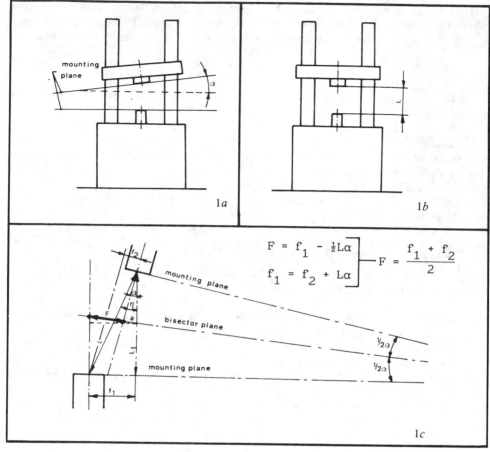

FIG. 1—*Schematic presentation of some applied definitions.*

2.4. Determination of the Alignment Error f_l at L

The piston rod is adjusted in the midstroke position and $L \leq 300$ mm. A rotating table or substitute is placed on the lower mounting plane. The shaft of a dial gauge touches the accurately finished measuring rim of the upper mounting plane. Now the crosshead is locked and the dial gauge is zeroed. The device is rotated 180° and the difference in values of the dial gauges is registered.

This measurement has to be repeated in directions perpendicular to each other (*A* and *B*) and at, for example, $L = 300$ mm. The alignment errors are $f_{lA(300)}$ and $f_{lB(300)}$, which is the difference of the corresponding readout values divided by two (Fig. 2).

2.5. Determination of the Angular Deflection α at L

The angle between the upper and lower mounting planes is measured with the instruments mentioned in Section 2.4. The piston rod is in the midstroke position and $L \leq 300$ mm. The dial gauge shaft rests against the upper mounting plane near the edge and is

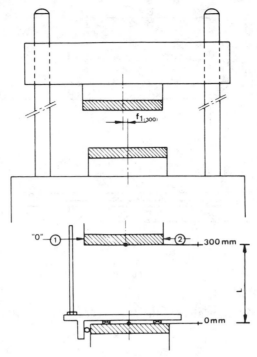

FIG. 2—*Setup for the measurement of* $f_{1(300)}$.

zeroed. The device is rotated 180° and the dial gauge is registered. Again, the measurement has to be repeated in directions perpendicular to each other, A and B.

At, for example, $L = 300$ mm, the difference of the corresponding readout values divided by D (Fig. 3) are the angular deflections $\alpha_{A(300)}$ and $\alpha_{B(300)}$ (tan $\alpha = \alpha$ if $\alpha \ll 1$).

The maximum angular deflection $\alpha_{(300)}$ can be calculated with

$$\alpha_{(300)} = \sqrt{\alpha^2_{A(300)} + \alpha^2_{B(300)}}$$

The direction (ψ_1) of $\alpha_{(300)}$ relative to the B-axis is calculated with:

$$\tan \psi_1 = \frac{\alpha_{A(300)}}{\alpha_{B(300)}}$$

The eccentricity in, for example, direction A is

$$F_{A(300)} = f_{1A(300)} - \tfrac{1}{2}L\alpha_{A(300)}$$

and is similar for direction B.

The maximum eccentricity (F_{300}) is

$$F_{(300)} = \sqrt{F^2_{A(300)} + F^2_{B(300)}}$$

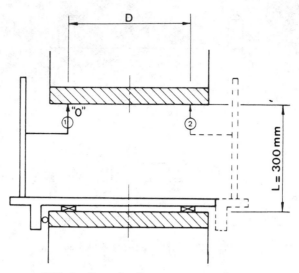

FIG. 3—*Setup for the measurement of* $\alpha_{(300)}$.

The direction (ψ_2) of $F_{(300)}$ relative to the B-axis is calculated with

$$\tan \psi_2 = \frac{F_{A(300)}}{F_{B(300)}}$$

2.6. Determination of the Alignment Error f_l at $L > 300$ mm $(f_{l(L)})$

The alignment error can be measured with the aid of an optical test setup as shown schematically in Fig. 4.

Before adjusting the alignment telescope, the crosshead is locked in a position as high as possible and the target is positioned. The measurements start when the crosshead is slightly lowered from the highest position and locked.

Every following measuring level will be done at a *lower* position of the crosshead until, for example, $L = 300$ mm has been reached. At every measuring level the alignment telescope and the clinometers (see section 2.7) should be readout at the same time.

When the relative measurements at $L = 300$ mm have been finished, the plan parallel plate with the telescope has to be removed and the absolute measurements, as described in sections 2.4 and 2.5, with the crosshead still locked in the same position have to be carried out. A correlation is then achieved between the relative and absolute measurements.

If there is no sufficient space, the telescope can usually be rotated over 90° instead of 180°. In this case the telescope tube must be mounted exactly parallel to the plan parallel plate. After this, the target must be positioned in such a way that by rotating the telescope the center of the target always is maintained in the center of the crosslines. At every measuring level two measurements will then be sufficient, namely in direction A, respectively in direction B. Two clinometers will be positioned in the same directions.

The absolute alignment errors at L are

$$f_{1A(L)} = f_{1A(300)} + \Delta f_{1A(L)}$$

and is similar for direction B.

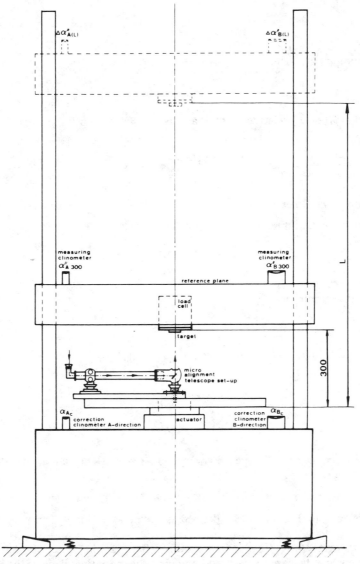

FIG. 4—*Setup for the determination of* $f_{I(L>300)}$ *and* $\alpha_{(L>300)}$.

2.7. Measurement of the Angular Deflection α at L > 300 mm ($\alpha_{(L)}$)

The angular deflection (α) is measured with two clinometers in the directions A and B located on the adjustable crosshead. To correct for angular variations of the complete machine while raising or lowering the crosshead, two reference clinometers are placed on the machine base in the directions A and B, respectively, and connected to each other to obtain differential readout values.

The readings of the clinometer setting at $L = 300$ mm are zeroed in Directions A and B. At $L = 300$ mm $+ \Delta L$, we find $\Delta\alpha_{A(L)}$ and $\Delta\alpha_{B(L)}$. Hence the angular deflection of the upper mounting plane relative to the lower one at L is

$$\alpha_{A(L)} = \alpha_{A(300)} + \Delta\alpha_{A(L)}$$

and is similar for direction B.

The maximum angular deflection $\alpha_{(L)}$ is

$$\alpha_{(L)} = \sqrt{\alpha_{A(L)}^2 + \alpha_{B(L)}^2}$$

The direction (ψ_1) of $\alpha_{(L)}$ relative to the B-axis is calculated with

$$\tan \psi_1 = \frac{\alpha_{A(L)}}{\alpha_{B(L)}}$$

2.8. Calculation of the Eccentricity $F_{(L)}$

The eccentricity at L in direction A can be calculated as

$$F_{A(L)} = f_{1A(L)} - \tfrac{1}{2}L\alpha_{A(L)}$$

and is similar for direction B.

The maximum eccentricity $F_{(L)}$ is

$$F_{(L)} = \sqrt{F_{A(L)}^2 + F_{B(L)}^2}$$

The direction (ψ_2) of $F_{(L)}$ relative to the B-axis is calculated with

$$\tan \psi_2 = \frac{F_{A(L)}}{F_{B(L)}}$$

3. Method to Determine the Alignment Error f_v and Angular Deflection α_v as a Result of Piston Rod Displacement

3.1. Definition Necessary to Determine the Alignment Errors f_v and α_v

The terms relating to these errors are defined as:

1. *Alignment error* f_v *(mm)* (Fig. 5). The position change of the center of the piston rod mounting plane when displaced measured parallel to the mounting plane of the load cell.
2. *Angular deflection* α_v *(radians)* (Fig. 5). The rotation of the mounting plane of the piston rod from the highest to the lowest position.

3.2. Measurement to Determine the Alignment Error f_v and Angular Deflection α_v with a Dial Gauge

The crosshead should be locked in such a position that the highest and lowest position of the piston rod can be measured.

The results at the highest position are

$$f_{v1A} \text{ and } \alpha_{v1A}$$

and are similar for direction B.

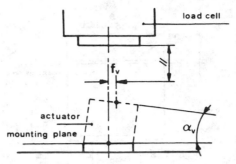

FIG. 5—*Schematic presentation of the definitions of the alignment error* (f$_v$) *and angular deflection* (α_v).

After lowering the piston to its lowest position the results are

$$f_{v2A} \text{ and } \alpha_{v2A}$$

and are similar for direction B.
    The alignment error in direction A is

$$f_{vA} = f_{v2A} - f_{v1A}$$

and is similar for direction B.
    The maximum alignment error is

$$f_v = \sqrt{f_{vA}^2 + f_{vB}^2}$$

The direction (θ_1) of f_v relative to the B-axis is calculated with

$$\tan \theta_1 = \frac{f_{vA}}{f_{vB}}$$

The angular deflection in direction A is

$$\alpha_{vA} = \alpha_{v2A} - \alpha_{v1A}$$

and is similar for direction B.
    The maximum angular deflection $\alpha_{(L)}$ is $\alpha_{(L)} = \sqrt{\alpha_{A(L)}^2 + \alpha_{B(L)}^2}$
    The direction (θ_1) of $\alpha_{(L)}$ relative to the B-axis is calculated with

$$\tan \psi_1 = \frac{\alpha_{A(L)}}{\alpha_{B(L)}}$$

4. Measurement to Evaluate the Alignment of the Loaded Frame

To check the alignment parameters of the load frame being loaded from zero to maximum load, a series of alignment specimens are designed and provided with strain gauges (Fig. 6). Opposite strain gauges in a section were arranged as a bending sensitive half bridge. Hence a tensile load in the specimen cannot be recognized by these strain gauge instruments. Clamping heads for these specimens are designed and tested as well.

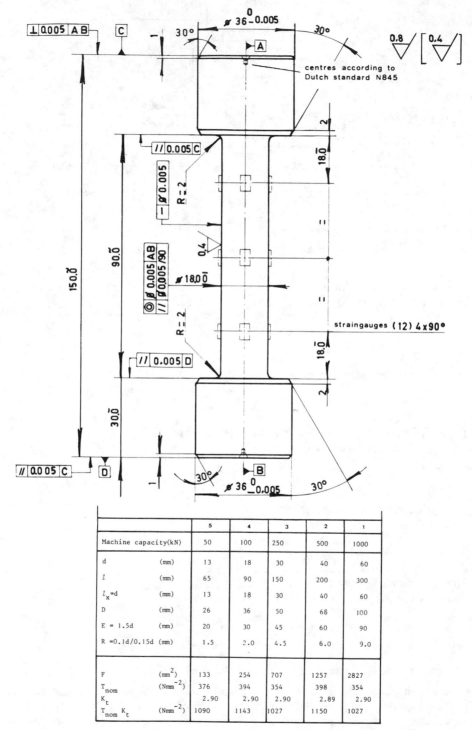

FIG. 6—*Dimensions of alignment specimen no. 4 and main dimensions for the series of alignment specimens as designed by the working party.*

		5	4	3	2	1
Machine capacity(kN)		50	100	250	500	1000
d	(mm)	13	18	30	40	60
l	(mm)	65	90	150	200	300
$l_x = d$	(mm)	13	18	30	40	60
D	(mm)	26	36	50	68	100
$E = 1.5d$	(mm)	20	30	45	60	90
$R = 0.1d/0.15d$	(mm)	1.5	2.0	4.5	6.0	9.0
F	(mm^2)	133	254	707	1257	2827
T_{nom}	(Nmm^{-2})	376	394	354	398	354
K_t		2.90	2.90	2.90	2.89	2.90
$T_{nom} K_t$	(Nmm^{-2})	1090	1143	1027	1150	1027

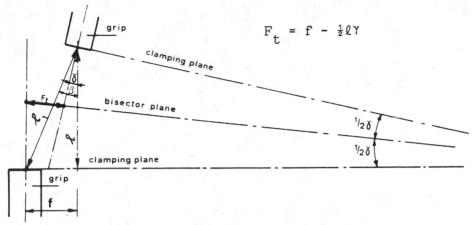

FIG. 7—*Schematic presentation of the definitions related to the alignment specimens.*

4.1. Definitions Necessary for the Alignment Specimens

Since clamping heads are introduced in the load frame, the definitions of the previously described errors should be adjusted. The related definitions now are:

1. *Clamping plane* (Fig. 7). Plane of the alignment specimen collar which is prestressed to clamp this specimen.
2. *Eccentricity of the load frame with clamping devices* F_t *(mm)* (Fig. 7). The distance between the intersections of the center normals of the upper and lower clamping planes with the bisector plane bisecting the angle between the clamping planes while the piston rod is in the midstroke position.
3. *Alignment error* f *(mm)* (Fig. 7). The distance between the center normal of the lower clamping plane and the center of the upper clamping plane while the piston rod is in the midstroke position.
4. *Angular deflection* γ *(radians)* (Fig. 7). The angle between the clamping planes while the piston rod is in the midstroke position.

4.2. Strains in the Alignment Specimen Due to Angular Deflection

According to the equations for the elastic deformation of a bar under an end load (Fig. 8), a linear relation exists between the strains $\Delta\varepsilon_d$, $\Delta\varepsilon_m$, and $\Delta\varepsilon_{l-d}$. Hence the results of the previous measurements can be compared.

Assuming that the axial load is zero, the strains induced by an alignment error f and an angular deflection γ between the clamping heads can simply be schematized in a bar or rod with diameter d and length l. As a result of f and γ in the upper clamping head, a sideload P_D and a bending moment M should behave, according to the equation

$$f = \frac{l^2}{EI}\left(\frac{P_D \cdot l}{3} - \frac{M}{2}\right) \rightarrow P_D = \frac{6EI}{l^2}\left(2\frac{f}{l} - \gamma\right) \tag{1}$$

$$\gamma = \frac{l}{EI}\left(\frac{P_D \cdot l}{2} - M\right) \rightarrow M = \frac{2EI}{l}\left(3\frac{f}{l} - 2\gamma\right) \tag{2}$$

The moment in a given cross section x is: $M_x = P_D \cdot x - M$.

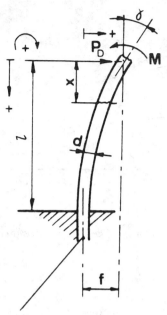

FIG. 8—*Schematic presentation of stresses in a rod induced by an alignment error* f *and angular deflection* γ *between the clamping heads.*

Then

$$\sigma_x = \frac{P_D \cdot x - M}{W} \rightarrow \varepsilon_x = \frac{P_D \cdot x - M}{EW}$$

$$\varepsilon_x = \frac{EI}{EW \cdot l} \left[6\frac{x}{l}\left(2\frac{f}{l} - \gamma\right) - 2\left(3\frac{f}{l} - 2\gamma\right) \right]$$

where $W = I/0.5d$.

$$\Delta\varepsilon_x = 2\varepsilon_x = \frac{d}{l} \left[6\frac{x}{l}\left(2\frac{f}{l} - \gamma\right) - 2\left(3\frac{f}{l} - 2\gamma\right) \right] \text{ (see Fig. 9)} \qquad (3)$$

$\Delta\varepsilon_x$ is the "output" of the bending sensitive half bridge corrected for strain gauge factor and wire resistance. With Eq 3, for any value of x the stresses in that particular cross section can be calculated for a given f, γ, d, and l (d and l are specimen diameter and specimen gauge length).

The strain gauges in the middle, for which $x = \frac{1}{2}l$, (Eq 3) yield

$$\Delta\varepsilon_{1/2l} = \Delta\varepsilon_m = \frac{d}{l} \left\{ \frac{\frac{1}{2}l \cdot 6}{l}\left(2\frac{f}{l} - \gamma\right) - 2\left(3\frac{f}{l} - 2\gamma\right) \right\} = \frac{d}{l}\gamma$$

In short

$$\Delta\varepsilon_m = \frac{d}{l} \cdot \gamma \qquad (4)$$

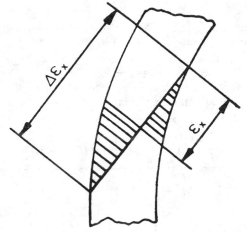

FIG. 9—*Strains in an alignment specimen's cross section.*

This means that *the strain in the middle of the specimen depends only on the angular deflection.*

From Eq 3 the alignment error (f) can be derived by substituting γ from Eq 4.

$$f = \frac{\Delta\varepsilon_x + 2\dfrac{d\gamma}{l}\left(3\dfrac{x}{l} - 2\right)}{6\dfrac{d}{l^2}\left(2\dfrac{x}{l} - 1\right)} \tag{5}$$

When $x = d$, then

$$f = \frac{\Delta\varepsilon_d + 2\Delta\varepsilon_m\left(3\dfrac{d}{l} - 2\right)}{6\dfrac{d}{l^2}2\left(\dfrac{d}{l} - 1\right)} \tag{5A}$$

With respect to the measuring accuracy, the minimum bending strain in the middle of the alignment specimen (ε_m) has to be at least 10 μ-strain, or, since $\Delta\varepsilon_m = 2\varepsilon_m$, the half bridge output is then 20 μ-strain.

With $\Delta\varepsilon_m = 20$ μ-strain and $\gamma = 10^{-4}$ rad (recommended in Chapter 6 of Ref 3), the dimensions l and d of the alignment specimen in relation to each other can be calculated with Eq 4.

This leads to

$$\frac{l}{d} = 5$$

For example, for a combination of the maximum angular deflection and eccentricity, the output of one of the other half bridges ($\Delta\varepsilon_d$ and $\Delta\varepsilon_{(l-d)}$) in the same (bending) plane as that related to the angular deflection will be larger.

4.3. The Eccentricity F_t

Based on the definitions as given in Section 4.1 and since γ and β are assumed to be very small: $l_1 = l$ and consequently

$$F_t = f - \tfrac{1}{2}l\gamma \tag{6}$$

Substitution of Eq 4 and Eq 5 in Eq 6 yields

$$F_t = \frac{\Delta\varepsilon_x + 2\Delta\varepsilon_m\left(3\dfrac{x}{l} - 2\right) - \dfrac{l^2}{2d}\Delta\varepsilon_m\left[6\dfrac{d}{l^2}\left(2\dfrac{x}{l} - 1\right)\right]}{6\dfrac{d}{l^2}\left(2\dfrac{x}{l} - 1\right)} = \frac{\Delta\varepsilon_x - \Delta\varepsilon_m}{6\dfrac{d}{l^2}\left(2\dfrac{x}{l} - 1\right)}$$

or, if $6(d/l^2)(2x/l - 1) = C_x$ and $x = d$ it becomes

$$F_t = \frac{\Delta\varepsilon_d - \Delta\varepsilon_m}{C_d} \tag{7}$$

where

$$C_d = 6\frac{d}{l^2}\left(2\frac{d}{l} - 1\right) \tag{8}$$

C_d is constant for alignment specimens with a dimension ratio $l/d = 5$ and with strain gauge bridges at distances $x = d$ and $x = \tfrac{1}{2}l$ from the upper clamping plane.

With Eqs 4 and 7, the angular deflection (γ) and the eccentricity (F_t) of the load frame with clamping heads can be determined in two directions perpendicular to each other.

From Eq 4 it follows

$$\gamma_A = \frac{l}{d}\Delta\epsilon_{m(A)} \text{ and } \gamma_B = \frac{l}{d}\Delta\epsilon_{m(B)}$$

The maximum angular deflection (γ) is

$$\gamma = \sqrt{\gamma_A^2 + \gamma_B^2}$$

The direction (δ_1) of γ relative to the B-axis is calculated with

$$\tan\delta_1 = \frac{\gamma_A}{\gamma_B}$$

With Eq 7, it follows consequently

$$F_{t(A)} = \frac{\Delta\epsilon_{d(A)} - \Delta\epsilon_{m(A)}}{C_d}$$

and is similar for direction B.

The maximum eccentricity is

$$F_t = \sqrt{F_{t(A)}^2 + F_{t(B)}^2}$$

The direction (δ_2), relative to the B-axis is calculated with

$$\tan \delta_2 = \frac{F_{t(A)}}{F_{t(B)}}$$

To determine the alignment errors, Eq 6 can be used, resulting in

$$f_A = F_{t(A)} + \tfrac{1}{2}l\gamma_{(A)}$$

and is similar for direction B.

The maximum alignment error is

$$f = \sqrt{f_A^2 + f_B^2}$$

The direction (δ_3) of f relative to the B-axis is calculated with

$$\tan \delta_3 = \frac{f_A}{f_B}$$

4.4. Proposal for a Series of Alignment Specimens

Based on the preceding sections, the working party proposes a series of alignment specimens with dimensions as given in Fig. 6 for the measurements in agreement with sections 4.1 and 4.2. The diameter of the specimens is such that at the nominal load of the fatigue testing machine the tensile stress is ≈ 400 MN/m^2. It is recommended to manufacture the specimens from a type of steel with $\sigma_{0.2} \geqq 1400$ MN/m^2 in view of the stress concentration of ~ 3 in the radius of the shoulder. The specimens have to be ground to the correct dimensions as specified for specimen 4 in Fig. 6.

With the alignment specimens including adequate clamping heads, it is possible to assess the effect of alignment error and angular deflection of the loaded frame.

4.5. Application of the Proposed Specimens

The influences of the angular deflection (γ) and the eccentricity (F_t), which the frame including the clamping devices bears upon the alignment specimen, are determined with the aid of strain gauges bonded to the specimen. If the angular deflection and the eccentricity are measured in this way, they should not be considered in a self-evident way to be equivalent with the errors of the unloaded frame.

The specimen can, among other things, be used for:

1. Comparative measurements, for instance, during periodical servicing of the machine.
2. The determination of the directions in which the angular deflection and eccentricity are effective on the specimen.
3. Comparative measurements between the unloaded and loaded condition of the machine.

After clamping the alignment specimen in the clamping devices, the output of the strain gauges in directions A and B can be read out at zero load.

With these data, calculations according to section 4.3 can be made of:

1. The maximum angular deflection γ.
2. The direction (δ_1) of the maximum angular deflection (γ) relative to the B-axis.
3. The eccentricity F_t of the upper clamping plane with respect to the lower clamping plane.
4. The direction (δ_2) of the maximum eccentricity (F_t) relative to the B-axis.
5. The alignment error f of the upper clamping plane relative to the lower clamping plane.
6. The direction (δ_3) of the maximum alignment error (f) relative to the B-axis.

By interpreting the results of these calculations, proposals for corrections of the load frame can be made.

Experiences using the proposed measuring methods and calculations are obtained by P. E. van Riesen [4].

5. Principles of Clamping an Alignment Specimen

To design a justifiable clamping device for an alignment specimen, account has to be taken of some criteria:

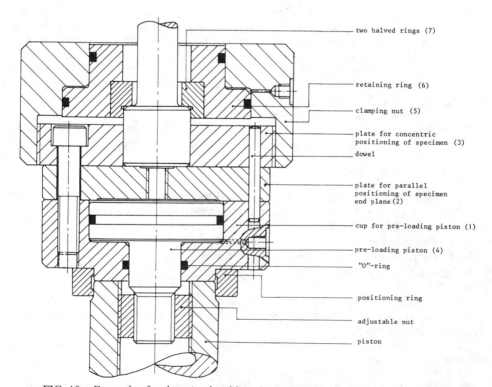

two halved rings (7)

retaining ring (6)

clamping nut (5)

plate for concentric positioning of specimen (3)

dowel

plate for parallel positioning of specimen end plane (2)

cup for pre-loading piston (1)

pre-loading piston (4)

"0"-ring

positioning ring

adjustable nut

piston

FIG. 10—*Example of a clamping head based on the presence of a threaded central hole in both piston rod and load cell.*

1. The mounting planes of the piston rod and of the load cell have to satisfy the requirements of section 2.1.

2. The clamping planes have to be parallel with the mounting plane of piston rod and load cell, respectively.

3. The load pressing the clamping planes of the alignment specimen has to be evenly distributed.

4. The magnitude of the load pressing the clamping planes has to be such that, while tensile loading to the nominal load of the specimen, an evenly distributed contact between end plane and clamping device is still preserved.

In Figs. 10 and 11 two designs are given which satisfy the listed criteria. The design of Fig. 10 is based on the presence of a threaded central hole in both piston rod and load cell, while that of Fig. 11 is based on a circular row of bolts to connect a clamping device.

Although the centering parts (2) and (3) could be made in one piece, this option was not selected in view of the expected difficulties to grind the planes parallel within the required limits as well as with regard to the coaxiality of the measuring/positioning rims.

The oil pressure necessary to obtain a sufficient preload to maintain a close contact between the axial planes has to be ≥ 332 bar for the 250-kN specimen and 257 bar for the 50-kN specimen. These rather high pressures can be obtained with simple hydraulic pumps because the required volume of fluid is negligible. It is also possible to use the pressure of the hydraulic system of the test machine itself. Because the oil pressure commonly applied in these machines is 250 or 210 bar, the diameter of the circular space has in that case to be increased accordingly.

Since many test machines are provided with hydraulic grips, it seemed worthwhile to

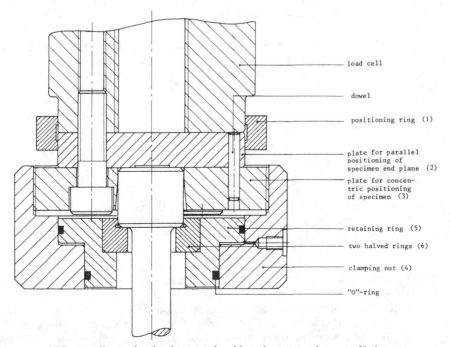

FIG. 11—*Example of a clamping head based on a circular row of bolts.*

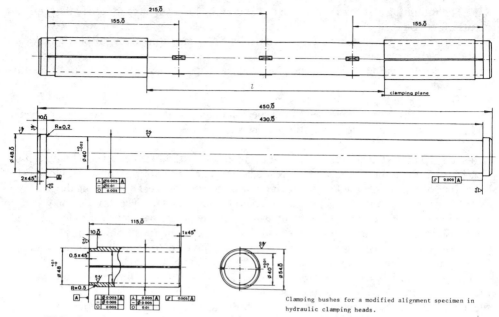

FIG. 12—*Alignment specimen and clamping bushes for a 500-kN load frame with hydraulic grips.*

also pay attention to this type of clamping. If a specification of the accuracy of the grips is available, a verification specimen as shown in Fig. 12 can be applied. Again the l/d ratio is 5.

To be able to reuse it, halved bushes protect the surface of the specimen. These bushes have to be rejected after one or two measurements being dented by the grips. It should be noted that the shown verification specimen can be used in tension to the nominal load of the machine.

6. Conclusions and Recommendations

1. Based on the specifications of the alignment accuracy offered by a number of manufacturers of testing machines, the following provisional requirements are proposed:

   1. Eccentricity, F at $0 < L \leq 300$ mm: ± 0.02 mm.
   2. Eccentricity, F at $300 < L <$ max.: ± 0.02 mm + 0.1 mm/m.
   3. Angular deflection, α at $0 < L \leq$ max.: $\leq 10^{-4}$ rad.
   4. Alignment error, f_v: ± 0.1 mm/m displacement.
   5. Angular deflection, α_v: $\leq 10^{-4}$ rad.

These proposed requirements are considered to be reasonable. Results obtained by using the described measuring methods have confirmed this.

2. From experience it appears that sufficient information can be obtained using only dial gauges and electronic clinometers and an alignment telescope to determine the alignment requirements. A good agreement exists between these results and the results obtained from the alignment specimen.

3. Application of these requirements is dependent of the objectives of the testing machine user. Nevertheless, it is strongly recommended to apply these whenever necessary.

4. After having given the opportunity to the manufacturers to accustom themselves to the proposed requirements, it seems reasonable to the working party to evaluate the effects after a certain period based on the experiences of users and manufacturers as well.

References

[1] Bressers, J., "Axiality of Loading," *Proceedings,* Symposium on Measurements of High Temperature Mechanical Properties of Materials, held at Teddington, United Kingdom, 3-5 June 1981, edited by M. S. Loveday et al., HMSO, London, pp. 278-294.

[2] Christ, B. W. and Swanson, S. R., "Alignment Problems in the Tensile Test," *Journal of Testing and Evaluation,* Vol. 4, No. 6, November 1976, pp. 405-417.

[3] Baten, R. J. H., d'Haen, H. J., Jacobs, F. A., Muller, M. K., Van Riesen, P.E., and Tjoa, G. L., "The Determination of the Permitted Size of the Alignment Error of a Load Frame and a Proposal for a Relevant Measuring Method," available on request from secretary, Working Party on Requirements for Fatigue Testing Machines, G. L. Tjoa, The Netherlands Energy Research Foundation, ECN, P.O. Box 1, 1755 ZG Petten, The Netherlands.

[4] Van Riesen, P. E., "Determination of the Alignment Errors of a 500-kN MTS Load Frame Model 311.21," Faculty of Mechanical Engineering, Dept. of Applied Mechanics, University Twente, The Netherlands, October 1987.

Roderich Fischer[1] and Erwin Haibach[2]

Checking and Improvement of the Alignment of Flat Specimen Gripping Devices

REFERENCE: Fischer, R. and Haibach, E., **"Checking and Improvement of the Alignment of Flat Specimen Gripping Devices,"** *Factors That Affect the Precision of Mechanical Tests, ASTM STP 1025,* R. Papirno and H. C. Weiss, Eds., American Society for Testing and Materials, Philadelphia, 1989, pp. 136–159.

ABSTRACT: The reliability of results of fatigue tests depends to a high degree on the accuracy of the alignment of the gripping devices used. Misalignments can result in undesired additional stresses and strains in the specimen under test and can lead to inaccurate test results. A method is presented by means of which the alignment of flat specimen grips can be checked and steps for its improvement derived using a specific set of strains, which are measured at eight selected points of a dummy specimen clamped in its four possible gripping positions. The mathematical description of the strains in the clamped specimen, including those caused by a productional precurvature, leads to an analysis of the measured strains by means of an evaluation scheme which is derived from the obtained mathematical equations and tailor-made for the use of a computer. The results of the analysis allow judgment as to the accuracy of the alignment and the derivation of steps for its improvement.

KEY WORDS: gripping device, alignment

1. Introduction

In fatigue testing of axially loaded specimens, the precision of the test results will depend to a great extent on an accurate alignment of the gripping devices used to load the specimens. Misalignment of gripping devices may result in inaccurate test results, as misalignment is likely to cause additional stresses in the specimen under test due to bending deformations when gripping and/or loading the specimen, and those effects, which add locally to the average stress due to pure tension or compression load, are not considered by the usual way of direct stress computation. The problem may be only partially overcome by the use of self-adjusting gripping devices. Hence, if the reliability of test results shall be secured, checking and improvement of the alignment of the gripping devices used is an indispensable demand in up-to-date fatigue testing techniques.

Furthermore, the gripping device must be precisely and rigidly fixed to the attachment faces of the load frame and the stiffness of the frame must be sufficiently high in order to maintain the accurate alignment when test loads are applied to the specimen. And, finally, the proper alignment once achieved must not get lost when changing specimens as could, for instance, occur when moving the crosshead of the frame.

[1] Carl Schenck AG, Darmstadt, West Germany.
[2] Seilpruefstelle der WBK, D-4630 Bochum 1, West Germany.

In this paper a method for checking the alignment of flat specimen gripping devices is presented, intended for use in axial load fatigue testing.

The method makes use of a dummy specimen (DS) applied with strain gauges, the readings of which are analyzed by means of a specific evaluation procedure, which is derived on the supposition that the gripping device produces a fixed end support for the specimens, but allows the specimens to bend like a beam when not exactly straight or when misaligned grips produce bending moments. Compared with simpler procedures that may be imagined to serve the needs, however, the proposed method is particularly designed to allow for any misleading effects from a DS used for checking that is not exactly flat and straight. Moreover, the proposed method provides detailed information on how to stepwise improve the alignment, if necessary.

The basic idea to separate the effects of a deformed DS from the effects of misaligned grips is to use a DS applied with strain gauges near the gripping ends, where the bending moments from misaligned grips are largest, and to clamp it in the four different positions possible. Information on the influence of the stiffness of the load frame will be obtained by taking the strain gauge readings for zero load, a tension load, and a compression load. Proper analysis of these data will then provide a separation of the measured strains into components of different origin such as the eventually existing precurvature of the DS and the various types of deformation of the DS due to the various modes of misalignment of the grips with and without the influence of the deformation of the frame under load. On the other hand, analytical relations may be derived showing what strains will result from each mode of misalignment, and in reverse they may serve to specify what geometrical adjustments of the grips have to be made for their improvement.

2. Experimental Procedure

The experimental part of the procedure consists of strain measurements using a DS, the main dimensions (Length l, Width b, and thickness s) of which are equal to those of the actual test specimens. Then, after checking and possibly necessary improvement of the alignment of the gripping device, the testing machine can be used for the actual fatigue tests without further change. The DS is applied with eight strain gauges, characterized by the Arabic numerals 1 to 8, and positioned rather close to the end sections of the free test length (Fig. 1).

Readings of these strain gauges are taken for the free DS (zero reading) and for the four different positions (I to IV) in which it may be gripped (Fig. 2), considering a zero load, a tension load, a compression load, and again a zero reading for each of the four gripping positions. The load levels may be selected with reference to the planned testing loads.

While Numbers 1 to 8 of the strain gauges refer to their positions on the DS, Locations A to H denote the space-bound strain gauge positions relative to the grips. For all four gripping positions the correlation between the two denotations may be seen from Table 1, which forms the basis of the evaluation process described in Section 4. Therefore, for the correct application of the evaluation process and for the proper interpretation of the results obtained therefrom, it is essential to observe exactly the numbering and characterizing of the strain gauges and their space-bound locations in the different gripping positions.

The evaluation process starts with entering the eight strain values measured in each of the four gripping positions into a scheme corresponding to Table 1. The subsequent evaluation and correct interpretation of the measured values require a proper mathematical description of the strains of the DS, which is given in Section 3.

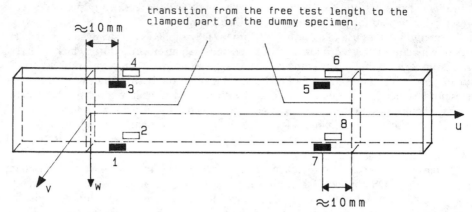

FIG. 1—*Position and numbering of the strain gauges applied to the dummy specimen.*

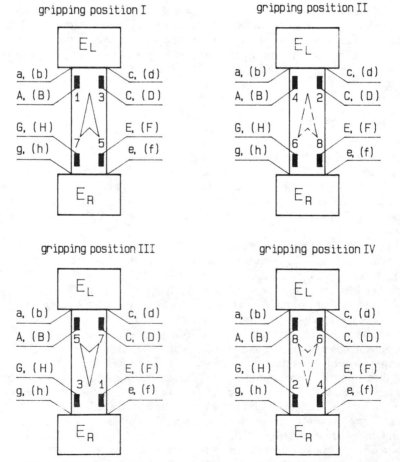

FIG. 2—*Gripping positions and space-bound locations of the strain gauges.*

TABLE 1—*Space-bound positions of the strain gauges 1 to 8 due to the different gripping positions.*

gripping position	strain gauge number							
	1	2	3	4	5	6	7	8
I	A	B	C	D	E	F	G	H
II	D	C	B	A	H	G	F	E
III	E	F	G	H	A	B	C	D
IV	H	G	F	E	D	C	B	A

3. Mathematical Description of the Strains

3.1 Basic Definitions

For the mathematical description of the strains of the DS, two Cartesian systems of coordinates are set up:

1. The specimen-bound system u, v, w (Fig. 1).
2. The space-bound system x, y, z (Fig. 3).

The x-axis of the latter is identical with the machine axis. The orientation of the two systems is selected in a way that they coincide if the DS is clamped in gripping position I.

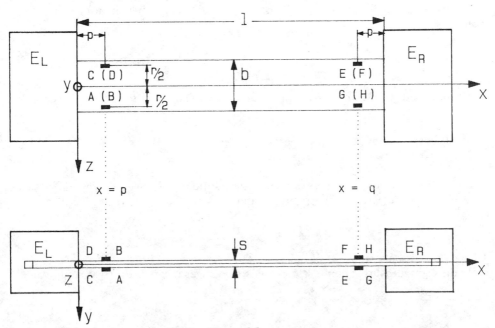

FIG. 3—*Clamped dummy specimen.*

The x position of the space-bound points A, B, C, and D is $x = p$; the x-position of E, F, G, H is $x = q$. The grip face leading edges which designate the transition from the free length to the clamped part of the DS are located at $x = 0$ and $x = l$.

Due to the various modes of misalignment, the clamped DS can undergo bending referred to the w-axis as well as to the v-axis and can be twisted around the u-axis. In addition, it must be assumed that the free DS has not an exactly prismatic shape but shows a productional precurvature, the compensation of which will cause strains. Furthermore, there is an average strain due to a tension or compression load. Therefore strains, which are measured by the strain gauges of the DS, can be considered to be composed of the following components, which are attributed to different causes:

1. Strain component ξ, caused by the compensation of a productional precurvature of the DS.
2. Strain component η, caused by bending referred to the w axis.
3. Strain component μ, caused by bending referred to the v axis,
4. Strain component ζ, caused by twisting around the u axis,
5. Strain component λ, caused by a longitudinal force.

Consequently, the total strain ε in the clamped DS can be described by

$$\varepsilon = \xi + \eta + \mu + \zeta + \lambda \tag{1}$$

The strain components η, μ, and ζ are obviously a measure of the accuracy of the alignment of the gripping device. Therefore, the objective of the whole procedure must be to separate these components from the measured total strain ε. To accomplish this, proper mathematical expressions for all single strain components of Eq 1 must be found which allow such an analysis. These expressions are derived in the following sections on condition that the simplifications of the elementary linear theory of strength of materials are valid.

3.2 Strain Component Caused by a Productional Precurvature

The assumed productional precurvature of the neutral plane of the DS may be described by the two functions:

1. $\kappa_1(u,w)$ = curvature in the planes w = constant.
2. $\kappa_2(u,w)$ = curvature in the planes u = constant.

If the DS is forced by proper external bending moments into an exactly prismatic shape, an arbitrary point (u,v,w) of it experiences a longitudinal strain which is described by

$$\xi(u,v,w) = v \cdot [\kappa_1(u,w) - v \cdot \kappa_2(u,w)] = v \cdot \varphi(u,w) \tag{2}$$

where v is Poisson's ratio.

At the eight measuring points 1 to 8, the strain gauges measure the eight strain values

$$\xi_i = \frac{s}{2} \cdot \varphi_i \text{ and } \xi_{i+1} = -\frac{s}{2} \cdot \varphi_i \tag{3}$$

where

$\varphi_i = \varphi(u_i, w_i)$, and

$i = 1, 3, 5, 7.$

Now, for all further reflections the DS can be considered as an exact prisma, which experiences the strains ξ_i or ξ_{i+1} at the eight measuring points.

3.3 Strain Component Caused by Bending Referred to the w-Axis

The strain component $\eta(x)$ in the clamped DS due to bending referred to the w-axis is given by

$$\eta(x) = v \cdot \frac{M_z(x)}{E \cdot I_w} \tag{4}$$

where

$M_z(x)$ = bending moment referred to z-axis in an arbitrary intersection x,
E = Young's modulus, and
I_w = area moment of inertia of the DS referred to its w-axis.

From Eq 4 follows that in the outermost fibers $v = +s/2$ the strain component η is positive if $M_z(x) > 0$.

The bending moment $M_z(x)$ is caused by the misalignment of the gripping device, which—in the most general case—is described by the dislocations f_L and f_R of the grip face leading edges from the machine axis and by the two angles α and β of the inclination of the gripping faces (Fig. 4). The balance of moments referred to the z-axis as shown in Fig. 4 leads to the moment function.

$$M_z(x) = -Q \cdot x + M_{z0}$$

where

Q = side load on DS, and
M_{z0} = boundary moment referred to the z-axis.

The relation between the bending moment $M_z(x)$ and the geometric quantities f_L, f_R, α, and β can be found by solving the differential equation of the elastic curve

$$E \cdot I_w \cdot \frac{d^2y}{dx^2} = -M_z(x) = Q \cdot x - M_{z0} \tag{5}$$

in consideration of the boundary conditions

$$\left(\frac{dy}{dx}\right)_{x=0} = \alpha, \quad \left(\frac{dy}{dx}\right)_{x=l} = \beta, \quad y(x=0) = f_L, \quad y(x=l) = f_R$$

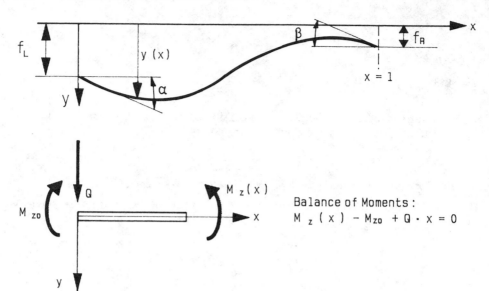

FIG. 4—*Elastic curve of the clamped dummy specimen.*

With this solution the strain in the outermost fibers, $v = +s/2$ and $v = -s/2$ is given by

$$\eta(x) = \pm s \left[\frac{3 \cdot f}{l^2}\left(1 - \frac{2 \cdot x}{l}\right) + \frac{2 \cdot \alpha}{l}\left(1 - \frac{3 \cdot x}{2 \cdot l}\right) + \frac{\beta}{l}\left(1 - \frac{3 \cdot x}{l}\right) \right] \qquad (6)$$

where

$$f = f_L - f_R \qquad (7)$$

From Eq 6 the strains caused by bending referred to the w-axis at the measuring points $x = p$ and $x = q$ of the DS can easily be derived

$$\eta(x = p) = \eta_p = \pm s \left[\frac{3 \cdot f}{l^2}\left(1 - \frac{2 \cdot p}{l}\right) + \frac{2 \cdot \alpha}{l}\left(1 - \frac{3 \cdot p}{2 \cdot l}\right) + \frac{\beta}{l}\left(1 - \frac{3 \cdot p}{l}\right) \right]$$
$$\eta_p = \pm s\Phi_N(p)$$
$$\eta(x = q) = \eta_q = \pm s \left[\frac{3 \cdot f}{l^2}\left(1 - \frac{2 \cdot q}{l}\right) + \frac{2 \cdot \alpha}{l}\left(1 - \frac{3 \cdot q}{2 \cdot l}\right) + \frac{\beta}{l}\left(1 - \frac{3 \cdot q}{l}\right) \right] \qquad (8)$$
$$\eta_q = \pm s\Phi_N(q)$$

The minus sign is valid for those strain gauges, which are assigned to the space-bound points B, D, F, and H. The subscript N of the function Φ defined by Eq 8 is to indicate that no longitudinal load is applied to the clamped DS.

If a tension force P is additionally applied to the clamped DS, the differential equation of the elastic curve

$$\frac{d^4y}{dx^4} - \rho^2 \cdot \frac{d^2y}{dx^2} = 0 \qquad (9)$$

with the abbreviation

$$\rho = \sqrt{\frac{P}{E \cdot I_w}}$$

follows from the equilibrium of a specimen element as shown in Fig. 5. The general solution of Eq 9 is

$$y = C_1 + C_2 x + C_3 \cdot \cosh(\rho x) + C_4 \cdot \sinh(\rho x)$$

where C_1, C_2, C_3, and C_4 are free constants of integration, which have to be determined by aid of the boundary conditions. Because of Eq 4 and the left-hand part of Eq 5, it is sufficient for the calculation of the strain $\eta(x)$ to determine only the constants C_3 and C_4. This can be accomplished without detailed consideration of the geometrical conditions, if it is assumed that the boundary moments

$$M_z(x = 0) = M_{z0} \text{ and } M_z(x = l) = M_{zl}$$

are applied by the gripping device to the clamped DS. Using these boundary conditions for the determination of C_3 and C_4, the strain in the outermost fibers $u = +s/2$ and $u = -s/2$ of the DS is given by

$$\eta(x) = \pm \frac{s}{2EI_w \cdot \sinh(\rho l)} \cdot \{M_{z0} \cdot \sinh[\rho(l - x)] + M_{zl} \cdot \sinh(\rho x)\} \qquad (10)$$

from which the strains at the measuring points follow by replacing x by p, respectively, q

$$\eta(x = p) = \eta_p = \pm \frac{s}{2EI_w \cdot \sinh(\rho l)} \cdot \{M_{z0} \cdot \sinh[\rho(l - p)] + M_{zl} \cdot \sinh(\rho p)\}$$

$$\eta_p = \pm s \cdot \Phi_T(p) \qquad (11)$$

$$\eta(x = q) = \eta_q = \pm \frac{s}{2EI_w \cdot \sinh(\rho l)} \cdot \{M_{z0} \cdot \sinh[\rho(l - q)] + M_{zl} \cdot \sinh(\rho q)\}$$

$$\eta_q = \pm s \cdot \Phi_T(q)$$

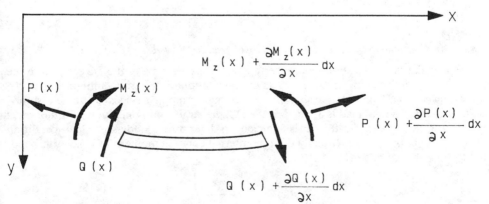

FIG. 5—*Forces and moments in the cross section areas of an element of the dummy specimen.*

The subscript T of the function Φ defined by Eq 11 is to indicate that a tension load is applied to the DS.

If a compression load is applied, the differential equation of the elastic curve changes from Eq 9 to

$$\frac{d^4y}{dx^4} + \rho^2 \frac{d^2y}{dx^2} = 0$$

Corresponding to Eqs 10 and 11, the equations

$$\eta(x) = \pm \frac{s}{2EI_w \cdot \sin(\rho l)} \cdot \{M_{z0} \cdot \sin[\rho(l - x)] + M_{zl} \cdot \sin(\rho x)\} \tag{12}$$

and

$$\eta(x = p) = \eta_p = \pm \frac{s}{2EI_w \cdot \sin(\rho l)} \cdot \{M_{z0} \cdot \sin[\rho(l - p)] + M_{zl} \cdot \sin(\rho p)\}$$

$$\eta_p = \pm s \cdot \Phi_c(p) \tag{13}$$

$$\eta(x = q) = \eta_q = \pm \frac{s}{2EI_w \cdot \sin(\rho l)} \cdot \{M_{z0} \cdot \sin[\rho(l - q)] + M_{zl} \cdot \sin(\rho q)\}$$

$$\eta_q = \pm s \cdot \Phi_c(q)$$

are obtained. The subscript C of the function Φ defined by Eq 13 indicates that a compression load is applied to the DS. From Eqs 8, 11, and 13 it follows that strain gauges 1 to 8 measure the strain components caused by bending referred to the w-axis

$$\eta_A = \eta_C = s \cdot \Phi(p) \qquad \eta_B = \eta_D = -s \cdot \Phi(p)$$
$$\eta_E = \eta_G = s \cdot \Phi(q) \qquad \eta_F = \eta_H = -s \cdot \Phi(q) \tag{14}$$

depending on which of the space-bound Points A to H they occupy in the different gripping positions I to IV. The subscripts N, T, and C of the function Φ, indicating the type of longitudinal load applied to the DS, can be omitted for further consideration because in the evaluation procedure, where use is made of Eq 14, these three loading conditions will be treated strictly separately.

3.4 Strain Components Caused by Bending Referred to the v-Axis

The calculation of the strain component μ caused by bending of the DS referred to the v-axis is accomplished in the same way as the calculation of the strain component η in the preceding section. Due to the design of flat specimen gripping devices, it can be assumed that, if no longitudinal load is applied to the DS, the strain component μ is of minor magnitude and importance. In that, it is sufficient to determine μ under zero load condition only as a function of the eventually existing internal bending moment $M_y(x)$. Corresponding to Eq 4 that results in

$$\mu(x) = \pm w \cdot \frac{M_y(x)}{E \cdot I_v} \tag{15}$$

from which follows that at measuring points $x = p$ and $x = q$ the strain component μ takes the values

$$\mu(x = p) = \mu_p = \pm r \cdot \frac{M_y(p)}{2EI_v} = \pm r\Theta_N(p)$$

$$\mu(x = q) = \mu_q = \pm r \cdot \frac{M_y(q)}{2EI_v} = \pm r\Theta_N(q)$$

(16)

where

I_v = area moment of inertia referred to the v-axis.

If a longitudinal force P is additionally applied to the clamped DS, the following equations for the strain component μ at the measuring points are valid in analogy to Eqs 11 and 13.

For a tension load

$$\mu(x = p) = \mu_p = \pm \frac{r}{2EI_v \cdot \sinh(\rho_1 l)} \cdot \{M_{y0} \cdot \sinh[\rho_1(l - p)] + M_{yl} \cdot \sinh(\rho_1 p)]$$

$$\mu_p = \pm r \cdot \Theta_T(p)$$

(17)

$$\mu(x = q) = \mu_q = \pm \frac{r}{2EI_v \cdot \sinh(\rho_1 l)} \cdot \{M_{y0} \cdot \sinh[\rho_1(l - q)] + M_{yl} \cdot \sinh(\rho_1 q)\}$$

$$\mu_q = \pm r \cdot \Theta_T(q)$$

For a compression load

$$\mu(x = p) = \mu_p = \pm \frac{r}{2EI_v \cdot \sin(\rho_1 l)} \cdot \{M_{y0} \cdot \sin[\rho_1(l - p)] + M_{yl} \cdot \sin(\rho_1 p)\}$$

$$\mu_p = \pm r \cdot \Theta_C(p)$$

(18)

$$\mu(x = q) = \mu_q = \pm \frac{r}{2EI_v \cdot \sin(\rho_1 l)} \cdot \{M_{y0} \cdot \sin[\rho_1(l - q)] + M_{yl} \cdot \sin(\rho_1 q)\}$$

$$\mu_q = \pm r \cdot \Theta_C(q)$$

where M_{y0} and M_{yl} are the specific boundary moments and ρ_1 is defined by

$$\rho_1 = \sqrt{\frac{P}{E \cdot I_v}}$$

Depending on which of the eight space-bound Points A to H the strain gauges occupy in the different gripping positions of the DS, they measure the following values of the strain component μ:

$$\mu_A = \mu_B = r \cdot \Theta(p) \qquad \mu_C = \mu_D = -r \cdot \Theta(p)$$
$$\mu_E = \mu_F = -r \cdot \Theta(q) \qquad \mu_G = \mu_H = r \cdot \Theta(q)$$

(19)

For the omitted subscripts N, T, and C of the function Θ, the comment to Eq 14 is valid, too.

3.5 Strain Component Caused by Twisting Around the u-*Axis*

By a small positive turn of the grip head E_R around the x-axis (Fig. 3), the neutral plane of the clamped DS experiences a curvature in the planes z = constant, which may be described by the function $\kappa_3(x,z)$. As, according to Section 3.2, the DS can be assumed to have an exactly prismatic shape, the function $\kappa_3(x,z)$ is symmetric with respect to the central point of the DS. Therefore κ_3 has the same magnitude ψ at each of the eight measuring points 1 to 8. Referring to the eight space-bound Points A to H, it is obvious that

$$\kappa_A = \kappa_B = \kappa_E = \kappa_F = -\psi \quad \text{and} \quad \kappa_C = \kappa_D = \kappa_G = \kappa_H = \psi$$

Then the values of the strain component ζ, which are measured by the strain gauges in the space-bound positions A to H are given by

$$\zeta_A = \zeta_E = -\frac{s}{2}(-\psi) = \frac{s}{2}\cdot\psi, \quad \zeta_B = \zeta_F = -(-\psi)\cdot\left(-\frac{s}{2}\right) = -\frac{s}{2}\cdot\psi$$
$$\zeta_C = \zeta_G = -\frac{s}{2}(\psi) = -\frac{s}{2}\cdot\psi, \quad \zeta_D = \zeta_H = -(\psi)\cdot\left(-\frac{s}{2}\right) = \frac{s}{2}\cdot\psi \tag{20}$$

if the relation between curvature of the neutral plane and strain as already used in Eq 2 is considered for uniaxial strains.

3.6 Strain Component Caused by a Longitudinal Force

By a pure longitudinal load with the magnitude P a uniform strain

$$\lambda = \pm\frac{P}{E\cdot A} \tag{21}$$

is produced in the DS with the cross-section Area A.

The minus sign in Eq 21 is assigned to a compression load. As the strains are measured in each of the four gripping positions j (j = I,II,III,IV), the longitudinal load P has to be adjusted anew after clamping the DS in the respective gripping position. Because of the limited accuracy of load adjustment, the actual load applied in each gripping position j is bound to differ by a small error ΔP_j from the exact value P. Therefore, in each gripping position j the real strain component λ_j, which is measured by the eight strain gauges, is given by

$$\lambda_j = \pm\frac{P + \Delta P_j}{E\cdot A} \quad (j = \text{I, II, III, IV}) \tag{22}$$

It is necessary to take the adjustment error of the longitudinal load into consideration because the strain error due to this adjustment error can be of the same order of magnitude as the strain components η, μ, and ζ. Therefore, in the evaluation of the measuring results, an attempt must be made to eliminate this error.

4. Evaluation of the Results of the Measurement

The total strain ε, which is measured at the eighth measuring points 1 to 8, and which is—according to Eq 1—composed of different components described by Eqs 3, 14, 19, 20, and 22, is a function of the measuring point itself and of the gripping position. For this reason, Eq 1 can be converted by the aid of Eqs 3, 14, 19, 20, and 22 to the form

$$\varepsilon_{ji} = \xi_i + \eta_{\delta(j,i)} + \mu_{\delta(j,i)} + \zeta_{\delta(j,i)} \pm \frac{F + \Delta F_j}{E \cdot A}$$

$$\varepsilon_{j,i+1} = \xi_{i+1} + \eta_{\delta(j,i+1)} + \mu_{\delta(j,i+1)} + \zeta_{\delta(j,i+1)} \pm \frac{F + \Delta F_j}{E \cdot A}$$

(23)

where i and $i + 1$ ($i = 1,3,5,7$) indicate the specimen-bound measuring points, j ($j = $ I,II,III,IV) indicates the gripping position, and $\delta(j,i)$ or $\delta(j,i + 1)$ represent the letter in Line j and Column i or $i + 1$ of Table 1, designating one of the space-bound locations A to H of the strain gauges of the clamped DS.

As already mentioned in Section 3.1, the objective of the evaluation procedure must be to separate the strain components η, μ, and ζ from the total strains ε_{ji} or ε_{ji+1} because these components are a measure of the accuracy of the alignment of the gripping device.

For this purpose the measured total strains are entered into a matrix scheme corresponding to Table 1 (Evaluation Scheme 1 on Table 2) and equated to the right side of Eq 23. The result of this procedure is shown in the Lines I, II, III and IV of the Evaluation Scheme 1, Table 2. From the expressions in these lines it can be seen that the functions Φ and Θ are eliminated by forming the average values

$$M_i = \frac{\varepsilon_{Ii} + \varepsilon_{IIi}}{2} \quad \text{and} \quad N_i = \frac{\varepsilon_{IIIi} + \varepsilon_{IVi}}{2} \quad (i = 1, 2. \ldots 8)$$

(24)

which are entered into the two lower lines of Evaluation Scheme 1, Table 2. The structure of M_i and N_i puts obstacles in the way to get one of the quantities φ or ψ isolated. To overcome these the following assumption can be made: According to Eq 2, the quantity φ is determined by the curvatures κ_1 and κ_2 of the neutral plane of the DS. If the DS is manufactured carefully, the magnitudes of, for example, κ_1 at the measuring points 1 and 3 are approximately equal and the signs are equal. This is also valid for κ_2 at the measuring points 1 and 3. The same assumption can be made for the measuring points 5 and 7, so that for further considerations the relations

$$\varphi_1 \approx \varphi_3, \mathrm{sgn}\varphi_1 = \mathrm{sgn}\varphi_3 \quad \text{and} \quad \varphi_5 \approx \varphi_7, \mathrm{sgn}\varphi_5 = \mathrm{sgn}\varphi_7$$

(25)

can be assumed to be valid. This leads to the formation of the expressions

$$T_1 = \tfrac{1}{8} [M_1 - M_2 - M_3 + M_4 + N_1 - N_2 - N_3 + N_4] = \frac{s}{2} \cdot \psi - \frac{s}{4}(\varphi_1 - \varphi_3)$$

$$T_2 = \tfrac{1}{8} [M_5 - M_6 - M_7 + M_8 + N_5 - N_6 - N_7 + N_8] = \frac{s}{2} \cdot \psi - \frac{s}{4}(\varphi_5 - \varphi_7)$$

the average value

$$T = \frac{T_1 + T_2}{2} = \frac{s}{2} \cdot \psi - \frac{s}{8}(\varphi_1 - \varphi_3 + \varphi_5 - \varphi_6) \approx \frac{s}{2}\psi$$

(26)

TABLE 2—*Evaluation Scheme 1.*

gripping positions	Specimen-bound measuring points			
	1	2	3	4
I	$\varepsilon_{I1} = \frac{s}{2}\varphi_1 + \frac{s}{2}\psi$ $+r\theta(p)+s\Phi(p)$ $+\frac{F+\Delta F_I}{EA}$	$\varepsilon_{I2} = -\frac{s}{2}\varphi_1 - \frac{s}{2}\psi$ $+r\theta(p)-s\Phi(p)$ $+\frac{F+\Delta F_I}{EA}$	$\varepsilon_{I3} = \frac{s}{2}\varphi_3 - \frac{s}{2}\psi$ $-r\theta(p)+s\Phi(p)$ $+\frac{F+\Delta F_I}{EA}$	$\varepsilon_{I4} = -\frac{s}{2}\varphi_3 + \frac{s}{2}\psi$ $-r\theta(p)-s\Phi(p)$ $+\frac{F+\Delta F_I}{EA}$
II	$\varepsilon_{II1} = \frac{s}{2}\varphi_1 + \frac{s}{2}\psi$ $-r\theta(p)-s\Phi(p)$ $+\frac{F+\Delta F_{II}}{EA}$	$\varepsilon_{II2} = -\frac{s}{2}\varphi_1 - \frac{s}{2}\psi$ $-r\theta(p)+s\Phi(p)$ $+\frac{F+\Delta F_{II}}{EA}$	$\varepsilon_{II3} = \frac{s}{2}\varphi_3 - \frac{s}{2}\psi$ $+r\theta(p)-s\Phi(p)$ $+\frac{F+\Delta F_{II}}{EA}$	$\varepsilon_{II4} = -\frac{s}{2}\varphi_3 + \frac{s}{2}\psi$ $+r\theta(p)+s\Phi(p)$ $+\frac{F+\Delta F_{II}}{EA}$
III	$\varepsilon_{III1} = \frac{s}{2}\varphi_1 + \frac{s}{2}\psi$ $-r\theta(q)+s\Phi(q)$ $+\frac{F+\Delta F_{III}}{EA}$	$\varepsilon_{III2} = -\frac{s}{2}\varphi_1 - \frac{s}{2}\psi$ $-r\theta(q)-s\Phi(q)$ $+\frac{F+\Delta F_{III}}{EA}$	$\varepsilon_{III3} = \frac{s}{2}\varphi_3 - \frac{s}{2}\psi$ $+r\theta(q)+s\Phi(q)$ $+\frac{F+\Delta F_{III}}{EA}$	$\varepsilon_{III4} = -\frac{s}{2}\varphi_3 + \frac{s}{2}\psi$ $+r\theta(q)-s\Phi(q)$ $+\frac{F+\Delta F_{III}}{EA}$
IV	$\varepsilon_{IV1} = \frac{s}{2}\varphi_1 + \frac{s}{2}\psi$ $+r\theta(q)-s\Phi(q)$ $+\frac{F+\Delta F_{IV}}{EA}$	$\varepsilon_{IV2} = -\frac{s}{2}\varphi_1 - \frac{s}{2}\psi$ $+r\theta(q)+s\Phi(q)$ $+\frac{F+\Delta F_{IV}}{EA}$	$\varepsilon_{IV3} = \frac{s}{2}\varphi_3 - \frac{s}{2}\psi$ $-r\theta(q)-s\Phi(q)$ $+\frac{F+\Delta F_{IV}}{EA}$	$\varepsilon_{IV4} = -\frac{s}{2}\varphi_3 + \frac{s}{2}\psi$ $-r\theta(q)+s\Phi(q)$ $+\frac{F+\Delta F_{IV}}{EA}$
M_i	$M_1 = \frac{s}{2}\varphi_1 + \frac{s}{2}\psi$ $+\frac{F}{EA}$ $+\frac{\Delta F_I + \Delta F_{II}}{2EA}$	$M_2 = -\frac{s}{2}\varphi_1 - \frac{s}{2}\psi$ $+\frac{F}{EA}$ $+\frac{\Delta F_I + \Delta F_{II}}{2EA}$	$M_3 = \frac{s}{2}\varphi_3 - \frac{s}{2}\psi$ $+\frac{F}{EA}$ $+\frac{\Delta F_I + \Delta F_{II}}{2EA}$	$M_4 = -\frac{s}{2}\varphi_3 + \frac{s}{2}\psi$ $+\frac{F}{EA}$ $+\frac{\Delta F_I + \Delta F_{II}}{2EA}$
N_i	$N_1 = \frac{s}{2}\varphi_1 + \frac{s}{2}\psi$ $+\frac{F}{EA}$ $+\frac{\Delta F_{III} + \Delta F_{IV}}{2EA}$	$N_2 = -\frac{s}{2}\varphi_1 - \frac{s}{2}\psi$ $+\frac{F}{EA}$ $+\frac{\Delta F_{III} + \Delta F_{IV}}{2EA}$	$N_3 = \frac{s}{2}\varphi_3 - \frac{s}{2}\psi$ $+\frac{F}{EA}$ $+\frac{\Delta F_{III} + \Delta F_{IV}}{2EA}$	$N_4 = -\frac{s}{2}\varphi_3 + \frac{s}{2}\psi$ $+\frac{F}{EA}$ $+\frac{\Delta F_{III} + \Delta F_{IV}}{2EA}$

of which is—because of Eq 25—an approximate value of the strain component ζ, which is caused exclusively by twisting the DS around its u-axis. Regarding Eq 26, from Eq 20 follows

$$\zeta_A = \zeta_D = \zeta_E = \zeta_H \approx T \quad \text{and} \quad \zeta_B = \zeta_C = \zeta_F = \zeta_G \approx -T \tag{27}$$

which is a first result of the isolation procedure.

TABLE 2—*Continued.*

Specimen-bound measuring points			
5	6	7	8
$\varepsilon_{I5} = \frac{s}{2}\varphi_5 + \frac{s}{2}\psi$ $-r\theta(q)+s\Phi(q)$ $+\frac{F+\Delta F_I}{EA}$	$\varepsilon_{I6} = -\frac{s}{2}\varphi_5 - \frac{s}{2}\psi$ $-r\theta(q)-s\Phi(q)$ $+\frac{F+\Delta F_I}{EA}$	$\varepsilon_{I7} = \frac{s}{2}\varphi_7 - \frac{s}{2}\psi$ $+r\theta(q)+s\Phi(q)$ $+\frac{F+\Delta F_I}{EA}$	$\varepsilon_{I8} = -\frac{s}{2}\varphi_7 + \frac{s}{2}\psi$ $+r\theta(q)-s\Phi(q)$ $+\frac{F+\Delta F_I}{EA}$
$\varepsilon_{II5} = \frac{s}{2}\varphi_5 + \frac{s}{2}\psi$ $+r\theta(q)-s\Phi(q)$ $+\frac{F+\Delta F_{II}}{EA}$	$\varepsilon_{II6} = -\frac{s}{2}\varphi_5 - \frac{s}{2}\psi$ $+r\theta(q)+s\Phi(q)$ $+\frac{F+\Delta F_{II}}{EA}$	$\varepsilon_{II7} = \frac{s}{2}\varphi_7 - \frac{s}{2}\psi$ $-r\theta(q)-s\Phi(q)$ $+\frac{F+\Delta F_{II}}{EA}$	$\varepsilon_{II8} = -\frac{s}{2}\varphi_7 + \frac{s}{2}\psi$ $-r\theta(q)+s\Phi(q)$ $+\frac{F+\Delta F_{II}}{EA}$
$\varepsilon_{III5} = \frac{s}{2}\varphi_5 + \frac{s}{2}\psi$ $+r\theta(p)+s\Phi(p)$ $+\frac{F+\Delta F_{III}}{EA}$	$\varepsilon_{III6} = -\frac{s}{2}\varphi_5 - \frac{s}{2}\psi$ $+r\theta(p)-s\Phi(p)$ $+\frac{F+\Delta F_{III}}{EA}$	$\varepsilon_{III7} = \frac{s}{2}\varphi_7 - \frac{s}{2}\psi$ $-r\theta(p)+s\Phi(p)$ $+\frac{F+\Delta F_{III}}{EA}$	$\varepsilon_{III8} = -\frac{s}{2}\varphi_7 + \frac{s}{2}\psi$ $-r\theta(p)-s\Phi(p)$ $+\frac{F+\Delta F_{III}}{EA}$
$\varepsilon_{IV5} = \frac{s}{2}\varphi_5 + \frac{s}{2}\psi$ $-r\theta(p)-s\Phi(p)$ $+\frac{F+\Delta F_{IV}}{EA}$	$\varepsilon_{IV6} = -\frac{s}{2}\varphi_5 - \frac{s}{2}\psi$ $-r\theta(p)+s\Phi(p)$ $+\frac{F+\Delta F_{IV}}{EA}$	$\varepsilon_{IV7} = \frac{s}{2}\varphi_7 - \frac{s}{2}\psi$ $+r\theta(p)-s\Phi(p)$ $+\frac{F+\Delta F_{IV}}{EA}$	$\varepsilon_{IV8} = -\frac{s}{2}\varphi_7 + \frac{s}{2}\psi$ $+r\theta(p)+s\Phi(p)$ $+\frac{F+\Delta F_{IV}}{EA}$
$M_5 = \frac{s}{2}\varphi_5 + \frac{s}{2}\psi$ $+\frac{F}{EA}$ $+\frac{\Delta F_I + \Delta F_{II}}{2EA}$	$M_6 = -\frac{s}{2}\varphi_5 - \frac{s}{2}\psi$ $+\frac{F}{EA}$ $+\frac{\Delta F_I + \Delta F_{II}}{2EA}$	$M_7 = \frac{s}{2}\varphi_7 - \frac{s}{2}\psi$ $+\frac{F}{EA}$ $+\frac{\Delta F_I + \Delta F_{II}}{2EA}$	$M_8 = -\frac{s}{2}\varphi_7 + \frac{s}{2}\psi$ $+\frac{F}{EA}$ $+\frac{\Delta F_I + \Delta F_{II}}{2EA}$
$N_5 = \frac{s}{2}\varphi_5 + \frac{s}{2}\psi$ $+\frac{F}{EA}$ $+\frac{\Delta F_{III} + \Delta F_{IV}}{2EA}$	$N_6 = -\frac{s}{2}\varphi_5 - \frac{s}{2}\psi$ $+\frac{F}{EA}$ $+\frac{\Delta F_{III} + \Delta F_{IV}}{2EA}$	$N_7 = \frac{s}{2}\varphi_7 - \frac{s}{2}\psi$ $+\frac{F}{EA}$ $+\frac{\Delta F_{III} + \Delta F_{IV}}{2EA}$	$N_8 = -\frac{s}{2}\varphi_7 + \frac{s}{2}\psi$ $+\frac{F}{EA}$ $+\frac{\Delta F_{III} + \Delta F_{IV}}{2EA}$

In order to isolate the strain components η and μ, which are caused by bending referred to the v-axis and w-axis of the DS, the 32 expressions

$$\varepsilon_{Ii} - M_i, \quad \varepsilon_{IIi} - M_i, \quad \varepsilon_{IIIi} - N_i, \quad \text{and} \quad \varepsilon_{IVi} - N_i \ (i = 1, 2 \ldots 8) \qquad (28)$$

have to be derived from the values in Evaluation Scheme 1, Table 2, in a way that those expressions defined by Eq 28, which are assigned by ε_{ji} to the same space-bound Location

TABLE 3—*Evaluation Scheme 2.*

gripping positions	Space-bound positions of the strain-gauges			
	A	B	C	D
I	$\varepsilon_{I1} - M_1 =$ $+r\theta(p)+s\Phi(p)$ $+\dfrac{\Delta F_I - \Delta F_{II}}{2EA}$	$\varepsilon_{I2} - M_2 =$ $+r\theta(p)-s\Phi(p)$ $+\dfrac{\Delta F_I - \Delta F_{II}}{2EA}$	$\varepsilon_{I3} - M_3 =$ $-r\theta(p)+s\Phi(p)$ $+\dfrac{\Delta F_I - \Delta F_{II}}{2EA}$	$\varepsilon_{I4} - M_4 =$ $-r\theta(p)-s\Phi(p)$ $+\dfrac{\Delta F_I - \Delta F_{II}}{2EA}$
II	$\varepsilon_{II4} - M_4 =$ $+r\theta(p)+s\Phi(p)$ $-\dfrac{\Delta F_I - \Delta F_{II}}{2EA}$	$\varepsilon_{II3} - M_3 =$ $+r\theta(p)-s\Phi(p)$ $-\dfrac{\Delta F_I - \Delta F_{II}}{2EA}$	$\varepsilon_{II2} - M_2 =$ $-r\theta(p)+s\Phi(p)$ $-\dfrac{\Delta F_I - \Delta F_{II}}{2EA}$	$\varepsilon_{II1} - M_1 =$ $-r\theta(p)-s\Phi(p)$ $-\dfrac{\Delta F_I - \Delta F_{II}}{2EA}$
III	$\varepsilon_{III5} - N_5 =$ $+r\theta(p)+s\Phi(p)$ $+\dfrac{\Delta F_{III} - \Delta F_{IV}}{2EA}$	$\varepsilon_{III6} - N_6 =$ $+r\theta(p)-s\Phi(p)$ $+\dfrac{\Delta F_{III} - \Delta F_{IV}}{2EA}$	$\varepsilon_{III7} - N_7 =$ $-r\theta(p)+s\Phi(p)$ $+\dfrac{\Delta F_{III} - \Delta F_{IV}}{2EA}$	$\varepsilon_{III8} - N_8 =$ $-r\theta(p)-s\Phi(p)$ $+\dfrac{\Delta F_{III} - \Delta F_{IV}}{2EA}$
IV	$\varepsilon_{IV8} - N_8 =$ $+r\theta(p)+s\Phi(p)$ $-\dfrac{\Delta F_{III} - \Delta F_{IV}}{2EA}$	$\varepsilon_{IV7} - N_7 =$ $+r\theta(p)-s\Phi(p)$ $-\dfrac{\Delta F_{III} - \Delta F_{IV}}{2EA}$	$\varepsilon_{IV6} - N_6 =$ $-r\theta(p)+s\Phi(p)$ $-\dfrac{\Delta F_{III} - \Delta F_{IV}}{2EA}$	$\varepsilon_{IV5} - N_5 =$ $-r\theta(p)-s\Phi(p)$ $-\dfrac{\Delta F_{III} - \Delta F_{IV}}{2EA}$
$\frac{1}{4}\Sigma$	$S_A = r\theta(p)$ $+s\Phi(p)$	$S_B = r\theta(p)$ $-s\Phi(p)$	$S_C = -r\theta(p)$ $+s\Phi(p)$	$S_D = -r\theta(p)$ $-s\Phi(p)$

$$K = \frac{1}{4}\,(S_A - S_B + S_C - S_D) = s\Phi(p) = \eta_p$$

$$R = \frac{1}{4}\,(S_A + S_B - S_C - S_D) = r\theta(p) = \mu_p$$

A to *H*, are written in one column. The assignment of the single expressions of Eq 28 to the single positions of Evaluation Scheme 2, Table 3, can be accomplished with the aid of Table 1. In Table 1, for example, the positions (I.1), (II.4), (III.5), and (IV.8) are assigned to the space-bound Location A. Therefore Column A of Evaluation Scheme 2, Table 3, has to be filled with the expressions

$$\varepsilon_{I1} - M_1, \quad \varepsilon_{II4} - M_4, \quad \varepsilon_{III5} - N_5, \quad \text{and} \quad \varepsilon_{IV8} - N_8$$

After having written all 32 expressions defined by Eq 28 into Evaluation Scheme 2, Table 3, the average value of the expressions of each column is formed and written into the fifth line of the scheme. By the mathematical structure of these eight average values S_A, S_B. . .S_H, it can be known that an isolation of the strain components η and μ is possible by forming the expressions

TABLE 3—*Continued.*

	Space-bound positions of the strain-gauges		
E	F	G	H
$\varepsilon_{I5}\quad -M_5 =$ $-r\theta(q)+s\Phi(q)$ $+\dfrac{\Delta F_I\ -\Delta F_{II}}{2EA}$	$\varepsilon_{I6}\quad -M_6 =$ $-r\theta(q)-s\Phi(q)$ $+\dfrac{\Delta F_I\ -\Delta F_{II}}{2EA}$	$\varepsilon_{I7}\quad -M_7 =$ $+r\theta(q)+s\Phi(q)$ $+\dfrac{\Delta F_I\ -\Delta F_{II}}{2EA}$	$\varepsilon_{I8}\quad -M_8 =$ $+r\theta(q)-s\Phi(q)$ $+\dfrac{\Delta F_I\ -\Delta F_{II}}{2EA}$
$\varepsilon_{II8}\quad -M_8 =$ $-r\theta(q)+s\Phi(q)$ $-\dfrac{\Delta F_I\ -\Delta F_{II}}{2EA}$	$\varepsilon_{II7}\quad -M_7 =$ $-r\theta(q)-s\Phi(q)$ $-\dfrac{\Delta F_I\ -\Delta F_{II}}{2EA}$	$\varepsilon_{II6}\quad -M_6 =$ $+r\theta(q)+s\Phi(q)$ $-\dfrac{\Delta F_I\ -\Delta F_{II}}{2EA}$	$\varepsilon_{II5}\quad -M_5 =$ $+r\theta(q)-s\Phi(q)$ $-\dfrac{\Delta F_I\ -\Delta F_{II}}{2EA}$
$\varepsilon_{III1}-N_1 =$ $-r\theta(q)+s\Phi(q)$ $+\dfrac{\Delta F_{III}-\Delta F_{IV}}{2EA}$	$\varepsilon_{III2}-N_2 =$ $-r\theta(q)-s\Phi(q)$ $+\dfrac{\Delta F_{III}-\Delta F_{IV}}{2EA}$	$\varepsilon_{III3}-N_3 =$ $+r\theta(q)+s\Phi(q)$ $+\dfrac{\Delta F_{III}-\Delta F_{IV}}{2EA}$	$\varepsilon_{III4}-N_4 =$ $+r\theta(q)-s\Phi(q)$ $+\dfrac{\Delta F_{III}-\Delta F_{IV}}{2EA}$
$\varepsilon_{IV4}\ -N_4 =$ $-r\theta(q)+s\Phi(q)$ $-\dfrac{\Delta F_{III}-\Delta F_{IV}}{2EA}$	$\varepsilon_{IV3}\ -N_3 =$ $-r\theta(q)-s\Phi(q)$ $-\dfrac{\Delta F_{III}-\Delta F_{IV}}{2EA}$	$\varepsilon_{IV2}\ -N_2 =$ $+r\theta(q)+s\Phi(q)$ $-\dfrac{\Delta F_{III}-\Delta F_{IV}}{2EA}$	$\varepsilon_{IV1}\ -N_1 =$ $+r\theta(q)-s\Phi(q)$ $-\dfrac{\Delta F_{III}-\Delta F_{IV}}{2EA}$
$S_E\ =-r\theta(q)$ $\quad+s\Phi(q)$	$S_F\ =-r\theta(q)$ $\quad-s\Phi(q)$	$S_G\ =+r\theta(q)$ $\quad+s\Phi(q)$	$S_H\ =+r\theta(q)$ $\quad-s\Phi(q)$

$$L = \frac{1}{4}\,(S_E - S_F + S_G - S_H) = s\Phi(q) = \eta_q$$

$$S = \frac{1}{4}(-S_E - S_F + S_G + S_H) = r\theta(q) = \mu_q$$

$$K = \tfrac{1}{4}(S_A - S_B + S_C - S_D) = s \cdot \Phi(p), \qquad L = \tfrac{1}{4}(S_E - S_F + S_G - S_H) = s \cdot \Phi(q)$$
$$R = \tfrac{1}{4}(S_A + S_B - S_C - S_D) = r \cdot \Theta(p), \qquad S = \tfrac{1}{4}(-S_E - S_F + S_G + S_H) = r \cdot \Theta(q) \tag{29}$$

With this result the strain component η caused by bending referred to the w-axis can be found from Eq 14

$$\eta_A = \eta_C = K, \quad \eta_B = \eta_D = -K, \quad \eta_E = \eta_G = L, \quad \eta_F = \eta_H = -L \tag{30}$$

and the strain component μ caused by bending referred to the u-axis can be found from Eq 19

$$\mu_A = \mu_B = R, \quad \mu_C = \mu_D = -R, \quad \mu_E = \mu_F = -S, \quad \mu_G = \mu_H = S. \tag{31}$$

At this point the objective of the evaluation procedure is achieved. The strain components μ, η, and ζ, which are a measure of the accuracy of the alignment of the gripping device, are determined from the measured total strains.

In order to provide a check or verification of the measured total strains, the following quantities can be calculated from the Lines I to IV of Evaluation Scheme 1, Table 2.

$$M_{I} = \frac{1}{8}\sum_{i=1}^{8}\varepsilon_{Ii} = \frac{P + \Delta P_{I}}{E \cdot A} \qquad M_{II} = \frac{1}{8}\sum_{i=1}^{8}\varepsilon_{IIi} = \frac{P + \Delta P_{II}}{E \cdot A}$$

$$M_{III} = \frac{1}{8}\sum_{i=1}^{8}\varepsilon_{IIIi} = \frac{P + \Delta P_{III}}{E \cdot A} \qquad M_{IV} = \frac{1}{8}\sum_{i=1}^{8}\varepsilon_{IVi} = \frac{P + \Delta P_{IV}}{E \cdot A}$$

$$\overline{M} = \frac{1}{4}\sum_{j=1}^{IV}M_{j} = \frac{P}{E \cdot A} + \frac{\Delta P_{I} + \Delta P_{II} + \Delta P_{III} + \Delta P_{IV}}{4E \cdot A}$$

If no longitudinal load is applied to the DS, M_{I}, M_{II}, M_{III}, and M_{IV} must theoretically be zero. Practically, they will have the magnitude of the error of the measuring system. If a longitudinal load is applied to the DS, M_{I}, M_{II}, M_{III}, M_{IV}, and $\overline{M}$ show the average strain λ with the averaged error caused by the limited accuracy of the load adjustment.

5. Checking of the Accuracy of the Alignment

The alignment of a gripping device can be considered as sufficiently accurate if the following conditions are met:

1. The additional stresses caused by the misalignments of the gripping device at those points of the specimens where the failure is expected must not exceed certain limits, which have to be defined according to the general conditions of the test.
2. The additional stresses at the transition area from the free length to the clamped part of the specimen must not produce failures in this area.

In the following explanations the center of the specimen is considered to be the starting point of the fatigue crack. To be able to judge whether a gripping device meets the above conditions, the magnitudes of the additional stresses at the mentioned critical points $x = 0$, $x = l/2$, and $x = l$ of the specimen must be known (Fig. 3). They can be determined from the strain components at the space-bound points A to H, which were found by the evaluation procedure described in Section 4. For this purpose the relation

$$q = l - p \tag{32}$$

which is evident from Fig. 3 is introduced and the critical points in the transition area from the free length to the clamped part of the specimen, which are identical with the starting and end points of the free length of the outermost fibers, are named a to h in analogy to the space-bound points A and H as shown in Fig. 2.

Because of the linear and—with respect to the central point of the specimen—symmetric character of the strain component ζ, the additional strain at the critical points $x = 0$, $x = l/2$, and $x = l$ caused by twisting the specimen around the u-axis is given by the following equations, which can easily be derived from Eq 27

$$\zeta_m = 0 \quad \text{(center of the specimen)}$$

$$\zeta_a = \zeta_e = \zeta_d = \zeta_h \approx T \cdot \frac{l}{l - 2p} \cdot \frac{b}{r} = T^* \tag{34}$$

$$\zeta_b = \zeta_c = \zeta_f = \zeta_g \approx -T \cdot \frac{l}{l - 2p} \cdot \frac{b}{r} = -T^*$$

From Eq 6 it can be seen that the strain component η is a linear function of the coordinate x if no logitudinal force is applied to the DS. Regarding this, the following equations can be derived from the quantities K and L in Eq 30, which determine the additional strain caused by bending referred to the w-axis at the critical points of the specimen

$$x = 0: K_0 = \frac{K(l - p) - L \cdot p}{l - 2p}$$

$$x = l: L_0 = \frac{L(l - p) - K \cdot p}{l - 2p} \tag{35}$$

$$x = l/2: K_m = \frac{K + L}{2}$$

If an additional longitudinal load is applied to the DS, the corresponding equations are for a tension load

$$x = 0: K_0 = \frac{K \cdot \sinh[\rho(l - p)] - L \cdot \sinh(\rho p)}{\sinh[\rho(l - 2p)]}$$

$$x = l: L_0 = \frac{L \cdot \sinh[\rho(l - p)] - K \cdot \sinh(\rho p)}{\sinh[\rho(l - 2p)]} \tag{36}$$

$$x = l/2: K_m = \frac{(K + L) \cdot \sinh[\rho(l - 2p)/2]}{\sinh[\rho(l - 2p)]}$$

and for a compression load

$$x = 0: K_0 = \frac{K \cdot \sin[\rho(l - p)] - L \cdot \sin(\rho p)}{\sin[\rho(l - 2p)]}$$

$$x = l: L_0 = \frac{L \cdot \sin[\rho(l - p)] - K \cdot \sin(\rho p)}{\sin[\rho(l - 2p)]} \tag{37}$$

$$x = l/2: K_m = \frac{(K + L) \cdot \sin[\rho(l - p)/2]}{\sin[\rho(l - 2p)]}$$

which can be derived from Eqs 10 and 12 under consideration of Eqs 11, 13, and 30. In analogy to Eq 30, the strain components at the critical points a to h caused by bending referred to the w-axis are

$$\eta_a = \eta_c = K_0, \quad \eta_b = \eta_d = -K_0, \quad \eta_e = \eta_g = L_0, \quad \eta_f = \eta_h = -L_0 \tag{38}$$

The same reflections with respect to the strain component μ and the corresponding Eqs 15, 16, 17, 18, 19, and 31 result in the following equations to determine the additional strain caused by bending with respect to the v-axis at the critical points of the DS.

Without longitudinal load

$$x = 0: R_0 \quad = \frac{R(l - p) - Sp}{l - 2p}$$

$$x = l: S_0 \quad = \frac{S(l - p) - Rp}{l - 2p} \tag{39}$$

$$x = l/2: R_m = \frac{R + S}{2}$$

with tension load

$$x = 0: R_0 \quad = \frac{R \cdot \sinh[\rho_1(l - p)] - S \cdot \sinh(\rho_1 p)}{\sinh[\rho_1(l - 2p)]}$$

$$x = l: S_0 \quad = \frac{S \cdot \sinh[\rho_1(l - p)] - R \cdot \sinh(\rho_1 p)}{\sinh[\rho_1(l - 2p)]} \tag{40}$$

$$x = l/2: R_m = \frac{(R + S) \cdot \sinh[\rho_1(l - 2p)/2]}{\sinh[\rho_1(l - 2p)]}$$

and for a compression load

$$x = 0: R_0 \quad = \frac{R \cdot \sin[\rho_1(l - p)] - S \cdot \sin(\rho_1 p)}{\sin[\rho_1(l - 2p)]}$$

$$x = l: S_0 \quad = \frac{S \cdot \sin[\rho_1(l - p)] - R \cdot \sin(\rho_1 p)}{\sin[\rho_1(l - 2p)]} \tag{41}$$

$$x = l/2: R_m = \frac{(R + S) \cdot \sin[\rho_1(l - p)/2]}{\sin[\rho_1(l - 2p)]}$$

In the cases of $\rho \ll 1$ and $\rho_1 \ll 1$ the sine functions in Eqs 37 and 41 and the hyperbolic sine functions in Eqs 36 and 40 can be replaced without a measurable error by the first terms of their power series expansion. Then the Eqs 35 and 39 are valid for all three load conditions. The strain values R_0, S_0, and R_m defined by Eqs 39, 40, and 41 are valid for the fibers in a distance of $z = \pm r/2$ from the neutral fiber (Fig. 3). Therefore the strains at the Points a to h on the edge fibers caused by bending referred to the v-axis are in analogy to Eq 31

$$\mu_a = \mu_b = R_0 \cdot \frac{b}{r}, \quad \mu_c = \mu_d = -R_0 \cdot \frac{b}{r}, \quad \mu_e = \mu_f = -S_0 \cdot \frac{b}{r}, \quad \mu_g = \mu_h = S_0 \cdot \frac{b}{r} \tag{42}$$

Consequently from Eqs 34, 38, and 42 the total additional strain caused by the misalignments of the gripping device in the transition area from the free to the clamped part of the DS can be derived as follows

$$\varepsilon_a = T^* + K_0 + R_0 \cdot \frac{b}{r} \qquad \varepsilon_e = T^* + L_0 - S_0 \cdot \frac{b}{r}$$

$$\varepsilon_b = -T^* - K_0 + R_0 \cdot \frac{b}{r}, \qquad \varepsilon_f = -T^* - L_0 - S_0 \cdot \frac{b}{r}$$

$$\varepsilon_c = -T^* + K_0 - R_0 \cdot \frac{r}{b} \qquad \varepsilon_g = -T^* + L_0 + S_0 \cdot \frac{b}{r} \tag{43}$$

$$\varepsilon_d = T^* - K_0 - R_0 \cdot \frac{r}{b} \qquad \varepsilon_h = T^* - L_0 + S_0 \cdot \frac{b}{r}$$

In the center of the DS the additional strain on the surface, which contains the points A, C, E, and G is given by

$$\varepsilon_m = K_m + R_m \cdot \frac{2 \cdot z}{r} \qquad (44)$$

and on the opposite surface by

$$\bar{\varepsilon}_m = -K_m + R_m \cdot \frac{2 \cdot z}{r} \qquad (45)$$

By the aid of Eqs 44 and 45 the additional strain in the center of the DS can be calculated for arbitrary distances z from the neutral fiber, for example, also for the distance of the root of the notch in the case where the actual tests are to be performed with notched specimens.

By means of the additional strains at the critical points of the DS calculated from Eqs 43, 44, and 45, it can be decided whether the alignment of the gripping device is sufficiently accurate.

6. Iterative Improvement of the Alignment

If the strain values, which are found by Eqs 43, 44, and 45 are untolerably high, the alignment of the gripping device in question has to be improved. Since the strain component μ is only of minor importance compared with the strain components η and ζ, the proposed correctional actions are limited to the reduction of the strain components η and ζ.

From the definition of the quantity T by Eqs 26 and 27, it follows that the grip head E_R is twisted relative to the grip head E_L in the sense of a right-handed screw, if $T > 0$. Thus, from the sign of the quantity T follows the direction of the correctional turn of the grip head E_R or E_L. The magnitude of the turn must be found by trial and subsequent checking by measuring the additional strains in the DS. If necessary, the whole procedure has to be repeated.

The strain component η refers to three possible errors of the alignment which are characterized by the geometrical quantities α, β, and f (Eq 7, Fig. 4). In order to be able to correct these errors, α, β and f must be known. From Eqs 8 and 30 and with regard to Eq 32 only the two defining equations

$$K = s \left[\frac{3 \cdot f}{l^2} \cdot \frac{l - 2p}{l} + \frac{\alpha}{l} \cdot \frac{2l - 3p}{l} + \frac{\beta}{l} \cdot \frac{l - 3p}{l} \right]$$
$$L = -s \left[\frac{3 \cdot f}{l^2} \cdot \frac{l - 2p}{l} + \frac{\alpha}{l} \cdot \frac{l - 3p}{l} + \frac{\beta}{l} \cdot \frac{2l - 3p}{l} \right] \qquad (46)$$

for these three unknown quantities can be found. The quantities α, β, and f are the boundary values of the elastic curve of the clamped DS, and the strain in the DS is proportional to the second derivation of the elastic curve. From this it can be inferred that it is not possible to determine the quantities α, β, and f exclusively from the known strain in the DS. However, the determination of one of these quantities in another way needs considerable technical expenditure and efforts. Therefore the most efficient alternative to overcome these difficulties is to calculate the remaining two quantities after having assigned a reasonable value to the first one so that an iterative improvement of the alignments becomes possible. Reasonable means, also, to make adequate allowance for the specific design of the gripping device in question. The following procedure is proposed. Eq 46 can be transformed to

$$\alpha = \frac{l}{3s} \cdot \left[K \cdot \frac{2l - 3p}{l - 2p} + L \cdot \frac{l - 3p}{l - 2p} \right] - \frac{f}{l}$$

$$\beta = -\frac{l}{3s} \cdot \left[K \cdot \frac{l - 3p}{l - 2p} + L \cdot \frac{2l - 3p}{l - 2p} \right] - \frac{f}{l}$$

(47)

from which follows with regard to Eq 35

$$\alpha - \beta = \frac{l}{s}(K + L) = \frac{2l}{s} \cdot K_m$$

(48)

proving, that the strain component η will be eliminated in the center of the DS if $\alpha - \beta$ becomes zero.

The principal sketch of the geometrical conditions of the misaligned gripping device on Fig. 6 shows that the mutual dislocation of the two ports of the grip heads is composed of a real centering error f_z and a component $d(\alpha + \beta)$ depending on the inclination of the gripping faces, so that

$$f = f_z + d(\alpha + \beta)$$

is valid, where d is a rotating radius the magnitude of which is—depending on the design of the grip head—between zero and the overall length of the grip head. Assuming that $f_z \ll d(\alpha + \beta)$, a first approximate solution for f can be found from the equation

$$f_1 = d(\alpha + \beta)$$

by means of which Eq 47 can be transformed to

$$\alpha_1 = \frac{l}{3s} \cdot \left[K \cdot \left(\frac{2l - 3p}{l - 2p} - \frac{d \cdot l}{(l + 2d)(l - 2p)} \right) + L \cdot \left(\frac{l - 3p}{l - 2p} + \frac{d \cdot l}{(l + 2d)(l - 2p)} \right) \right]$$

$$\beta_1 = -\frac{l}{3s} \cdot \left[K \cdot \left(\frac{l - 3p}{l - 2p} + \frac{d \cdot l}{(l + 2d)(l - 2p)} \right) + L \cdot \left(\frac{2l - 3p}{l - 2p} - \frac{d \cdot l}{(l + 2d)(l - 2p)} \right) \right]$$

from which a first approximation α_1 and β_1 of α and β can be calculated. (Note that in this section the subscripts 1, 2, or general i do not refer to the strain gauges of the DS, but to the steps of an iterative improving procedure).

Now, the inclination of the gripping faces of the grip head E_L has to be changed mechanically by the correction angle γ_1, so that the new inclination $\alpha_2 = \alpha_1 + \gamma_1$ is achieved. By this the mutual dislocation of the two ports of the grip heads is changed, too, however, by the distance $t \cdot \gamma_1$, where t is a rotating radius, which depends on the kind of mechanical manipulation, and which is not necessarily the same as the one mentioned before. Corresponding to E_R, the inclination of E_L is changed by δ_1 and the dislocation by $t \cdot \delta_1$.

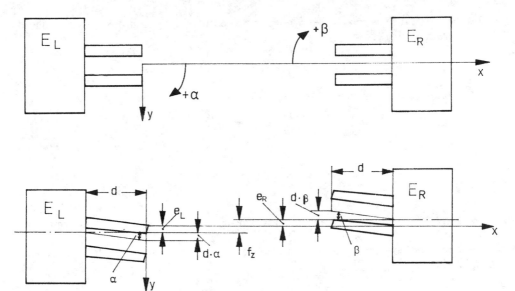

$$f_L = e_L + d \cdot \alpha \qquad f_Z = e_L + e_R \qquad f_R = -e_R - d \cdot \alpha$$

$$f = f_L - f_R = e_L + e_R + d\,(\alpha + \beta) = f_Z + d\,(\alpha + \beta)$$

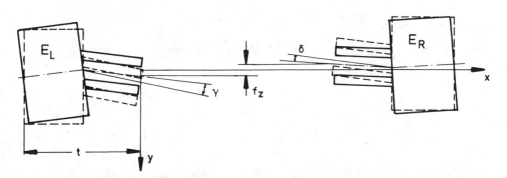

FIG. 6—*Gripping device:* (top) *accurately aligned;* (center) *misaligned;* (bottom) *corrected.*

The values of the correction angles γ_1 and δ_1 are determined by the following conditions:

1. The strain component η in the center of the DS must become zero. This is—according to Eq 44, 45, and 48—fulfilled if the difference between the corrected inclination angles $\alpha_2 = \alpha_1 + \gamma_1$ and $\beta_2 = \beta_1 + \delta_1$ becomes zero

$$\alpha_1 + \gamma_1 - (\beta_1 + \delta_1) = 0$$

2. The initial dislocation $d(\alpha_1 + \beta_1)$ of the two ports of the grip heads is to be reduced to the pure centering error:

$$d(\alpha_1 + \beta_1) + t(\gamma_1 + \delta_1) = 0$$

From these conditions the two defining equations

$$\gamma_1 = -\frac{\alpha_1}{2}\left(1 + \frac{d}{t}\right) + \frac{\beta_1}{2}\left(1 - \frac{d}{t}\right)$$

$$\delta_1 = \frac{\alpha_1}{2}\left(1 - \frac{d}{t}\right) - \frac{\beta_1}{2}\left(1 + \frac{d}{t}\right)$$

for γ_1 and δ_1 can be derived, by means of which the numerical values of γ_1 and δ_1 can be calculated.

After the first correction the gripping device is still subject to the inclinations of the gripping faces

$$\alpha_2 = \alpha_1 + \gamma_1 = \alpha_1 - \frac{\alpha_1}{2}\left(1 + \frac{d}{t}\right) + \frac{\beta_1}{2}\left(1 - \frac{d}{t}\right)$$

$$\beta_2 = \beta_1 + \delta_1 = \beta_1 + \frac{\alpha_1}{2}\left(1 - \frac{d}{t}\right) - \frac{\beta_1}{2}\left(1 + \frac{d}{t}\right)$$

and to the mutual dislocation of the ports of the grip heads

$$f_2 = f_1 + t(\gamma_1 + \delta_1) = f_1 - d(\alpha_1 + \beta_1)$$

Substituting α_2, β_2, and f_2 into Eq 46, the strain component η after the first correction step is found by

$$K_2 = \frac{K_1 - L_1}{2} \cdot \frac{1 - \dfrac{d}{t}}{1 + \dfrac{2d}{l}}, \qquad L_2 = -\frac{K_1 - L_l}{2} \cdot \frac{1 - \dfrac{d}{t}}{1 + \dfrac{2d}{l}}$$

After i repetitions of the correction procedure, the corresponding equations are

$$K_{i+1} = \frac{K_i - L_i}{2} \cdot \frac{1 - \dfrac{d}{t}}{1 + \dfrac{2d}{l}}, \qquad L_{i+1} = -\frac{K_i - L_i}{2} \cdot \frac{1 - \dfrac{d}{t}}{1 + \dfrac{2d}{l}}$$

These equations demonstrate:

1. After the first and each further correction K_i equals $-L_i$

$$K_i = -L_i \qquad (i = 1, 2 \ldots n)$$

This means because of the Eqs 30 and 35 that the strain component η is eliminated in the center of the DS.

2. If $0 < d \leq t$, then because of $t < 0$ the inequality

$$0 \leq \frac{1 - \dfrac{d}{t}}{1 + \dfrac{2 \cdot d}{l}} < 1$$

is valid, from which follows cogently

$$|K_{i+1}| < |K_i| \quad \text{and} \quad |L_{i+1}| < |L_i| \quad (i = 1,2 \ldots n)$$

This means, with each step of the correction, the strain component η in the DS decreases.

Acknowledgments

The method described in this paper was developed at the Fraunhofer-Institut für Betriebsfestigkeit (LBF) in Darmstadt, West Germany.

Donald W. Scavone[1]

Development of an Instrumented Device to Measure Fixture-Induced Bending in Pin-Loaded Specimen Trains

REFERENCE: Scavone, D. W., **Development of an Instrumented Device to Measure Fixture-Induced Bending in Pin-Loaded Specimen Trains,"** *Factors That Affect the Precision of Mechanical Tests, ASTM STP 1025,* R. Papirno and H. C. Weiss, Eds., American Society for Testing and Materials, Philadelphia, 1989, pp. 160–173.

ABSTRACT: Mechanical testing of various types is conducted using pinned clevis fixtures to transmit test loads to a range of specimen configurations. To ensure reliable repeatable results, fixture-induced load distribution variations must be kept to a minimum. This paper describes a simple, inexpensive, instrumented device and data analysis technique for measuring fixture-induced bending (that is, transverse nonuniformity of load application) of compact type [C(T)] specimen fixtures. Finite-element analysis was performed to illustrate the importance of uniform load distribution. The instrumented device may be used for other sizes or configurations of pin-loaded trains after consideration of appropriate dimensional modifications to the instrumented device.

KEY WORDS: bending, alignment, pinned clevis, compact specimen, stress intensity K

Mechanical testing of various types is conducted using pinned clevis fixtures to transmit test loads to a range of specimen configurations. To ensure reliable repeatable results, fixture-induced load distribution variations must be kept to a minimum. Maintaining load train straightness and concentricity provides some measure of alignment. These steps alone, however, do not account for the bending (that is, transverse nonuniformity of load application) contributions caused by machining and assembly of the various load train components. This paper describes a simple, inexpensive, instrumented device and data analysis technique for measuring fixture-induced bending of 0.4-in.-thick (0.4-T) compact type [C(T)] specimen fixtures. The technique may be used for other sizes or configurations of pin-loaded trains after consideration of appropriate dimensional modifications to the instrumented device.

To illustrate the importance of quantifying the load distribution in the pinned fixture, a pair of three-dimensional elastic finite-element analyses were performed to determine the variation in stress intensity factor, K, with variation in the load distribution. In the baseline analysis, the load was distributed uniformly among nine nodal points along the top inside surface of the loading pinhole. In the second analysis, nonuniform loading of the pinhole was simulated by applying a linear load distribution, which varied from 120% of the average on one side of the specimen to 80% on the other side. The commercially available finite-element code, ADINA[2], was used to perform the analyses. The energy release

---

[1] Senior specialist—Mechanics of Materials, General Electric Co., Knolls Atomic Power Laboratory, Schenectady, NY 12301-1072.
[2] ADINA Research and Development, Inc., 71 Elton Ave., Watertown, MA, 02172.

rate, G, was subsequently computed by the method of virtual crack extension, using a special purpose postprocessor called VIRTUAL, developed at the General Electric Company Corporate Research and Development Center [1]. The variation of the stress intensity factor, K, along the crack front (that is, through the specimen thickness) was computed from the energy release rate, G.

Figure 1 shows the ratio of the results from the two analyses. The results clearly indicate that the $\pm 20\%$ load variation produces a $\pm 8\%$ variation in G. This converts to a $\pm 4\%$ variation in the stress intensity factor, K. Since all of this is based upon linear analysis, the variation in K produced by other magnitudes of load variation may be determined by scaling these results in a linear fashion.

As will be demonstrated subsequently in this report, a 0.1016-mm difference in hole spacing on one side of a 0.4-T clevis relative to the other side will result in a 10% variation in load, and thereby a $\pm 2\%$ variation in stress intensity, along the crack front. A clevis hole spacing difference of this magnitude would result from drilling a 0.4-T clevis pinhole slightly more than 1° off from perpendicular.

Experimental Procedure

Device Design and Fabrication

Initial attempts at quantifying fixture-induced bending made use of a standard 0.4-T C(T) specimen modified to allow for the installation of strain gages as shown in Fig. 2. This specimen was installed in the load train, and individual strain readings were obtained over

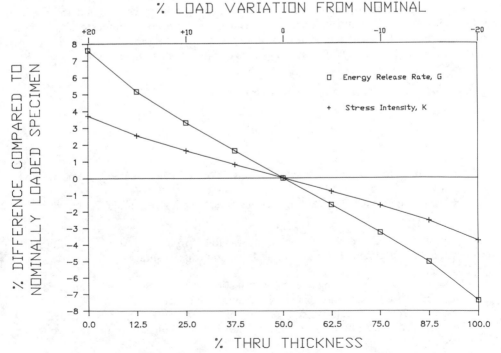

FIG. 1—*Results of finite-element analysis of a nonuniformly loaded specimen relative to a uniformly loaded specimen.*

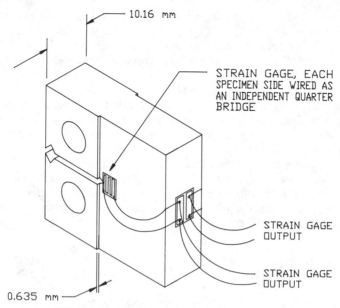

FIG. 2—*Modified compact type (C(T)) specimen.*

a typical load range. To account for inherent error of the instrumented specimen due to relative gage placement, specimen machining, etc., the specimen was rotated 180° about an axis normal to the machined notch and reinstalled in the train. A duplicate data set was then taken. Inherent specimen error contributions in each data set should be equal and opposite, thereby arithmetically cancelling. This technique is summarized by

$$\% \text{ Apparent Bending (For Either Orientation)} = \frac{Ea - Eb}{[(Ea + Eb)/2]} \times 100$$

% Bending = % Apparent Bending 0° Position + % Apparent Bending 180° Position

where

Ea = the strain measured by the gage on the side of the instrumented C(T) positioned near clevis Side A, and

Eb = the strain measured by the gage on the side of the instrumented C(T) positioned near clevis Side B.

In practice, data produced by this technique was found to be nonrepeatable and difficult to interpret. The specimen configuration itself has four disadvantages:

1. The typical C(T) machining tolerances in section thickness, hole placement, and hole parallelism are not tight enough for use as a calibration device.

2. Inherent error due to relative gage placement is exaggerated by the moment effect.

3. Low strain outputs at typical test loads result in low resolution in the relative strain measurements.

4. The configuration requires time-consuming multiple orientations to yield necessary data.

In view of these problems, the device described herein was developed. It consists of a pair of independent ligaments as shown in Fig. 3. In use, the two individual ligaments are installed simultaneously in a clevis as shown in Fig. 4. This device addresses each of the above concerns as follows:

1. Machining tolerances are tight (Fig. 5), with the ligaments being simultaneously jig ground as a matched set. To further ensure equivalent hole spacing of each ligament in a set, the set may be simultaneously overloaded on a common pin so as to cause yielding of the pinholes. (This step was not required for the ligament sets fabricated for this study. As-received dimensions from the simultaneous jig-grinding operation were well within the drawing-specified tolerances.)

2. Individual calibration of each ligament following strain gage installation eliminates the concerns of possible gage mismatch and/or gage placement differences (this will be discussed further in the following section, "Device Calibration"). Axial gage placement significantly reduces errors due to relative gage placement by eliminating the amplifying effect of the loading moment. Also, the two gages per ligament, wired in series as a quarter bridge (Fig. 6), electrically cancel any bending signal produced within an individual ligament.

3. The thin section size reduces inherent error from hole nonparallelism and results in high strain outputs for finer resolution.

4. The ability to independently calibrate each ligament of a set allows a single run to yield the necessary data to characterize a fixture.

The device is adaptable to any clevis and pin fixture by changing the pinhole size and spacing, and ligament thickness. The maximum possible ligament thickness is slightly less

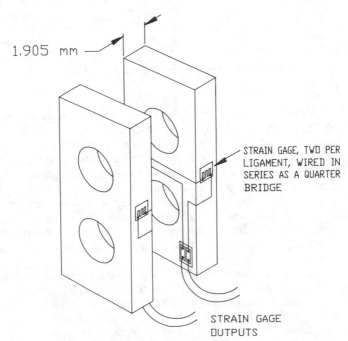

1.905 mm

STRAIN GAGE, TWO PER LIGAMENT, WIRED IN SERIES AS A QUARTER BRIDGE

STRAIN GAGE OUTPUTS

FIG. 3—*Ligament set.*

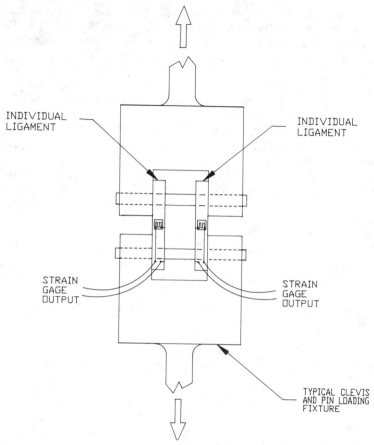

FIG. 4—*Ligament set installed in a clevis fixture.*

than one half the clevis opening (to preclude contact between the ligaments and thereby ensure independent ligament response). A thin cross section reduces errors due to ligament hole nonparallelism and also ensures ample strain within the ligament for readout resolution. Consideration must be given to the stress concentration factor at the ligament pinhole and to the fact that a given ligament will be subjected to more than 50% of the total train load in a train with bending. The optimum thickness for a specific application is influenced by the foreknowledge of the requirements of subsequent testing. For instance, load train alignment typically improves with increasing train load. If a clevis set is to be used for threshold-type fatigue testing, bending characteristics at low load levels is much more important than it would be for low-cycle fatigue (that is, high loads) testing.

The actual design of the loading ligaments for a particular application is the end product of many design criteria, both technical and practical. For example, during the course of the development of this technique, the load range of interest was 4.45 to 8.90 kN. A material with a high yield strength was chosen (age hardenable stainless steel, type 17-4 PH, condition H900, 0.2% offset yield strength 1172 MPa, Young's modulus of elasticity 28.5 ×

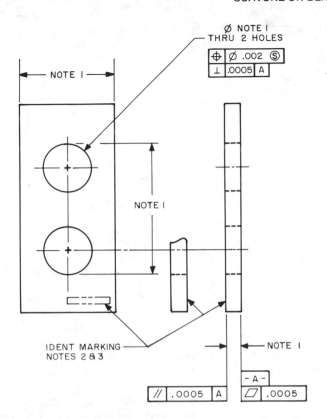

NOTES:
1. THIS DIMENSION SHALL NOT VARY BY MORE THAN .0005 BETWEEN ITEM 1 AND 2 IN A MATCHED SET.
2. IDENTIFICATION MARKING .10-.15 HIGH CHARACTERS, REMOVE UPSET MATERIAL AFTER MARKING.
3. TWO LOADING LIGAMENTS MAKE A MATCHED SET. MACHINING TO BE PERFORMED WHILE ITEMS 1 AND 2 ARE CLAMPED TOGETHER, WITH IDENT MARKINGS TOUCHING.
4. ALL INTERNAL AND EXTERNAL SURFACES SHALL BE 32/ FINISH.
5. TOLERANCES ARE IN INCHES, .002 in. = .051 mm

FIG. 5—*Ligament set critical machining tolerances.*

10^6 psi[3]) so as to allow as small a cross section as practical without resulting in failure at the pinholes. The cross section was chosen to be sufficient to prevent pinhole failure even if, due to nonuniform load distribution, one ligament was subjected to 150% of the nominal load per ligament. The ligament thickness was chosen to be as small as practical while still allowing for the installation of a narrow strain gage. After selecting the ligament material and thickness, the ligament width was made the minimum necessary which could be expected to sustain the desired maximum load, again without failing at the pinholes.

To summarize ligament thickness selection, the optimum thickness is the thinnest that

---

[3] 1 psi = 6.895 kPa.

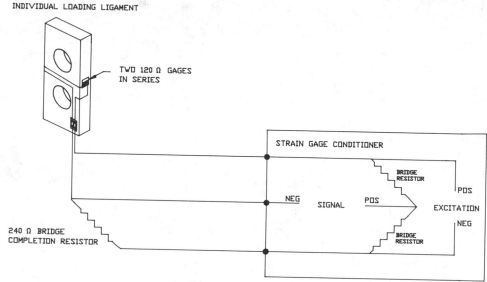

FIG. 6—*Ligament instrumentation schematic.*

will be safely within its elastic limit while still producing an ample strain signal when the train is loaded to the load level of interest.

Micro-Measurements[4] strain gage pattern 015DJ was selected for its small overall width (0.51 mm) of the grid and solder tab. However, the overall backing width of this pattern exceeded the loading ligament width. The excess backing beyond the actual grid width was trimmed prior to gage installation. The gages were centered on each ligament edge (Fig. 3), with the grid orientation parallel to the loading axis. The acceptance criteria for gage positioning is discussed in the following section.

Device Calibration

The intent of the precisely matched machining of a ligament pair, and the subsequent individual calibration of each ligament, is to produce two essentially identical load cells. Ideally, each ligament (that is, load cell) within a matched set is identical such that when the set is loaded on common pins in a perfectly aligned clevis fixture, the load measured by each ligament will be one half the total applied load. During the course of this research, the total load range over which test fixture loading uniformity was to be characterized was 4.45 to 8.90 kN. Each ligament was calibrated over a load range of one half the desired full train load.

Individual calibration of each ligament following strain gage installation eliminates the concerns of possible gage mismatch and/or gage placement differences. As the intent of the device is to serve as a load-measuring instrument, the indicated strain magnitudes, which would be affected by gage installation accuracy, are not of primary interest. The essential requirement is that each ligament be calibrated such that it can accurately indicate the magnitude of an unknown applied load. It is not essential that the strain indicated by each ligament be identical, so long as the load calibration data account for any differences. Therefore, the criteria used during calibration to determine acceptability of a ligament set strain gage installation is equivalent strain versus load slopes (that is, calibration factor)

---

[4] A division of Measurements Group, Inc., P.O. Box 27777, Raleigh, NC 27611.

within 2.0% for a matched set. However, as comparison of actual strains from ligament to ligament is also an indication of how precisely matched the machining of a set is, the additional criteria that no single strain versus load data point differ by more than 2.0% was applied.

Calibration was conducted on a screw-driven load frame. The calibration load train consisted of an upper universal joint, a 4.45-kN ring dynamometer (uncertainty in tension, 1.96 N) traceable to the United States National Bureau of Standards (NBS), and a modified 0.4-T compact-type specimen clevis of the flat-bottom hole configuration (as described in ASTM Test Method for Plane-Strain Fracture Toughness of Metallic Materials, E 399-83, Annex A4, Fig. A4.2). The modified 0.4-T clevis had an opening width of 2.54 mm to reduce pin bending during calibration of the thin ligaments.

After exercising the calibration load train to its full intended load of 4.45 kN, strain versus load data (Tables 1 and 2) was obtained in 0.44-kN increments throughout the calibration range. For each ligament, a linear regression analysis was performed on the average strain versus ring dynamometer indicated load from three repetitions. The individual regression line slopes matched within 1.6%. No single strain versus load data point differed between ligaments by more than 1.6%.

To establish load measurement uncertainty, ligament load was then back calculated from the average strain measurement using the slope and intercept data as follows

Ligament Load = (Ave. Strain-Ligament Intercept)/Ligament Slope

TABLE 1—*Individual ligament calibration data, Ligament A.*

Ring Load	Measured Strain				Back-Calculated Load	Residual, *res*
	1st Run	2nd Run	3rd Run	Average		
500	1289	1291	1290	1290	497.2	2.8
600	1539	1540	1540	1540	601.1	1.1
700	1782	1783	1782	1782	702.0	2.0
800	2022	2023	2023	2023	801.9	1.9
900	2257	2259	2258	2258	899.8	0.2
1000	2492	2496	2494	2494	998.0	2.0

Regression Analysis: X = ring dynamometer indicated load
Y = average strain

Constant (Y intercept)	94.49206
Standard error of Y estimate	5.469879
R squared	0.999881
No. of observations	6
Degrees of freedom	4
X coefficient	2.404380
Standard error of coefficient	0.013075

$$\text{RMS One Sigma Standard Deviation} = \sqrt{\left(\frac{(res_1)^2 + (res_2)^2 \cdots (res_n)^2}{n-1}\right) + RSD^2}$$

where:
residual, *res* = ring indicated load − ligament back-calculated load,
RSD = ring dynamometer standard deviation, and
ring dynamometer standard deviation = ring uncertainty/2.4.

NOTE: Loads are in pounds force, 1 pound force = 4.44822 newtons.

TABLE 2—*Individual ligament calibration data, Ligament B.*

Ring Load	Measured Strain				Back-Calculated Load	Residual, *res*
	1st Run	2nd Run	3rd Run	Average		
500	1273	1273	1273	1273	497.4	2.6
600	1518	1518	1518	1518	600.9	0.9
700	1757	1757	1757	1757	701.8	1.8
800	1994	1994	1994	1994	802.0	2.0
900	2226	2226	2226	2226	900.0	0.0
1000	2458	2458	2458	2458	997.9	2.1

Regression Analysis: X = ring dynamometer indicated load
Y = average strain

Constant (Y intercept)	95.42857
Standard error of Y estimate	5.156133
R squared	0.999891
No. of observations	6
Degrees of freedom	4
X coefficient	2.367428
Standard error of coefficient(s)	0.012325

$$\text{RMS One Sigma Standard Deviation} = \sqrt{\left(\frac{(res_1)^2 + (res_2)^2 \cdots (res_n)^2}{n-1}\right) + RSD^2}$$

where
residual, *res* = ring indicated load − ligament back-calculated load,
RSD = ring dynamometer standard deviation, and
ring dynamometer standard deviation = ring uncertainty/2.4.

NOTE: Loads are in pounds force, 1 pound force = 4.44822 newtons.

Root-mean-square (RMS) one sigma standard deviation was then computed on the back-calculated load data. The ring dynamometer standard deviation was included in this computation as an additional random variable. The resultant standard deviation was then multiplied by 2.4 to obtain uncertainty as defined by ASTM Practice for Calibration of Force Measuring Instruments for Verifying the Load Indication of Testing Machines (E 74-83, Paragraph 7.4). The resultant uncertainty of the ligaments used in this research was 21.8 N.

Device Usage

In use, with the two individual ligaments installed simultaneously in a 0.4-T clevis, the load train is exercised to its full intended load of 8.90 kN (twice the maximum load used in individual ligament calibration), and strain versus load data are obtained in 0.89-kN increments (0.44 kN per ligament nominal). Using the strain versus load calibration linear regression data for each ligament, it is then possible to compute directly the load on each clevis side as was described in the preceding calibration section

Ligament Load = (Strain-Ligament Intercept)/Ligament Slope

The sum of both ligament loads is a direct measure of the total train load. Ideally, this load would be uniformly shared by each ligament. Deviation of a ligament loading from nominally one half the total load is then a measure of fixture-induced bending.

Nominal Load per Ligament = Sum of Ligament Loads/2

$$\text{Ligament \% Bending} = \frac{\text{Ligament Load} - \text{Nominal Load per Ligament}}{\text{Nominal Load per Ligament}} \times 100$$

The maximum uncertainty in the measured percent bending will occur at the load range minimum. The load uncertainty of the individual ligaments is directly reflected in the load train bending measurement uncertainty. In this case, the 21.8-N ligament load uncertainty corresponds to a load train measured bending uncertainty of 0.5% at the 4.45-kN minimum train load.

To facilitate interpretation of the results as they relate to the fixture under test, one side of the clevis is designated the reference, and the magnitude and polarity of all bending measurements are recorded relative to the reference side. For example, consider a poorly machined clevis fixture loaded to 4.50-kN total load. With the clevis left side designated the reference, the measured applied load on the left side is 2.16 kN, and the right side 2.34 kN. The resulting ±4.0% load distribution from nominal would be reported as a bending value of −4.0%.

It must be understood that the device cannot pinpoint the specific cause, or causes, of poor loading uniformity in a particular load train. However, using the above convention, the necessary corrective action becomes more apparent.

Device Checkout

For checkout, the modified 0.4-T clevis and 4.45-kN ring dynamometer of the calibration load train were replaced with a standard 0.4-T clevis and 8.90-kN ring dynamometer. A ligament set was installed and data generated as described above. The data indicated the clevis loaded uniformly within ±0.3% throughout the load range 4.45 to 8.90 kN (2.22 to 4.45 kN per ligament).

Next, to stimulate bending conditions due to a poorly machined clevis, various size shims were inserted between a clevis hole and loading pin on one side only (Fig. 7). The ligament set was then loaded to 6.23 kN (nominally 3.115 kN per ligament, a value within each ligament's calibrated range) and bending magnitude and direction determined. The resulting data (Fig. 8) indicate an approximately linear response of measured bending to shim thickness. The device correctly indicated bending direction and demonstrated the ability to consistently resolve the bending difference resulting from 0.0127-mm change in shim thickness. (This experiment required the use of undersized pins to allow for shim installation. It was observed that the slope of a ligament's strain versus load data increases with decreasing pin diameter. Data linearity was not affected. Therefore, if actual loads on each ligament, rather than relative load distribution, are needed the device must be calibrated with the actual pin diameter to be used for fixture evaluation. In normal use of the device this is not a concern, as the standard fixture pin size will be used both for ligament calibration and for fixture evaluation.)

To demonstrate use of the device in correcting a clevis deficiency, the clevis nominal bending measurements were determined. One clevis side was then shimmed and bending quantified. An equivalent shim was then inserted between the same pin and clevis interface on the other side of the specimen. The measured bending returned to nominal. In actual use, the device would be used to indicate necessary corrective machining of the clevis.

To verify the ability of the device to quantify a fixture-induced bending condition, the fixture illustrated in Fig. 9 was devised. This fixture made possible the introduction of a repeatable bending force to the load train by means of a dead weight side load. For this experiment the clevis was again loaded to 6.23 kN and bending magnitude and direction

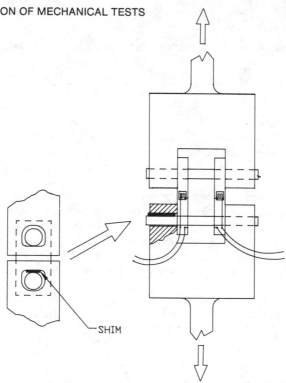

FIG. 7—*Ligament set checkout using a shim to induce nonuniform loading.*

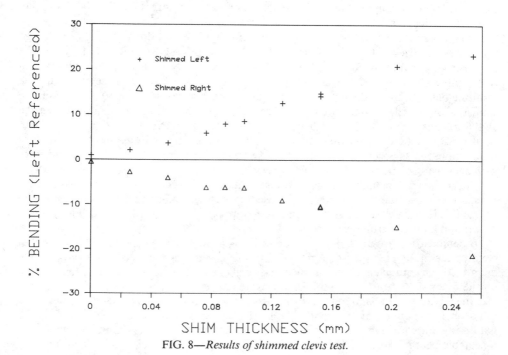

FIG. 8—*Results of shimmed clevis test.*

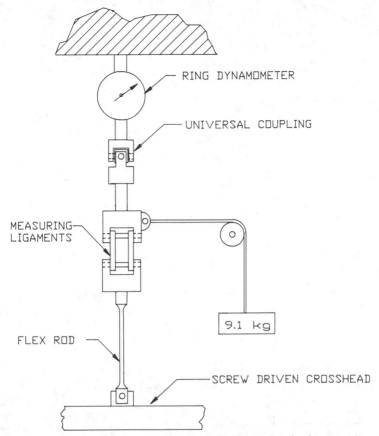

RING DYNAMOMETER

UNIVERSAL COUPLING

MEASURING LIGAMENTS

9.1 kg

FLEX ROD

SCREW DRIVEN CROSSHEAD

FIG. 9—*Ligament set checkout with force applied normal to load train axis to induce nonuniform loading.*

determined. A 88.96-N force was then applied normal to the clevis reference side and its effect on bending determined. The side load was then removed and bending measurements again taken to confirm the return of the train to the original condition. The device consistently measured the induced bending at −7.1% ±0.3%. Upon removal of the side load, bend measurements consistently returned to within 0.2% of original.

One objective of this design, relative to the modified C(T) approach, was to eliminate the need for a second data run following 180° specimen rotation. Following the experiment described above, the individual ligaments were transposed side to side. The results were identical to those above within ±0.1%. Therefore, it was concluded that the design met the objective and the 180° specimen rotation need not be performed.

Discussion

As an example of an actual use of the technique, the device was used to evaluate the design of a multiple specimen load train. The load frame was physically inaccessible and therefore required the load train be installed remotely. The train is dropped into the frame crosshead, via a crane, onto a beveled seat (Fig. 10). The design question was "Was the

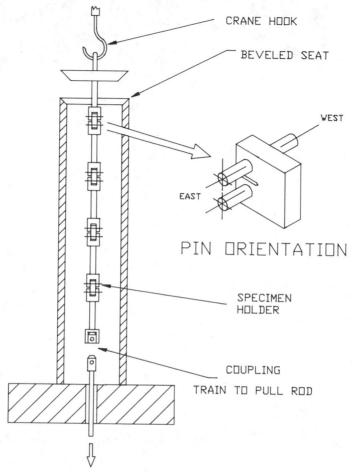

CRANE HOOK

BEVELED SEAT

WEST

EAST

PIN ORIENTATION

SPECIMEN HOLDER

COUPLING

TRAIN TO PULL ROD

FIG. 10—*Load train seating angle test fixture.*

beveled seat adequate to ensure no more than $\pm 5\%$ variation in load across the crack front of any of the individual test specimens?"

The instrumented device was installed in a load train specimen holder in place of a specimen. Specimen pin orientation was east to west. Unnotched specimen blanks were installed in the remaining positions. The load train then was deliberately seated on the beveled surface under worst-case conditions. The actual load distribution of the instrumented train position was then measured. This process was repeated for each of the remaining train positions. The seat design was proven to maintain $< \pm 5\%$ variation in all positions throughout the load range of interest and, therefore, required no redesign. Figure 11 illustrates the results of this investigation. As would be expected, north/south seat offsets (normal to pin orientation) produced negligible bending changes, while east/west seat offsets produced measurable bending influences.

One potential drawback of the device is handling difficulties associated with the two separate ligaments. A few approaches currently under consideration are a combination leaf spring spacer and handle, or potting the calibrated set in an elastic compound such as room temperature vulcanizing silicone rubber (RTV). Another approach would be to use the

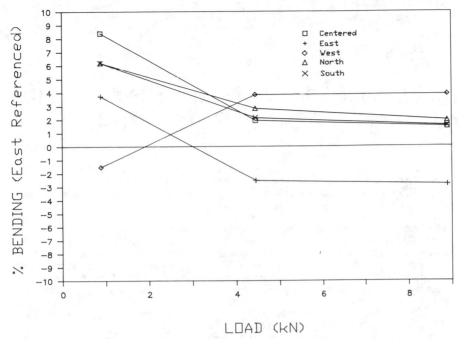

FIG. 11—*Results of load train seating angle test.*

device to characterize the loading accuracy of a specific load train. That load train, in turn, could then be used to calibrate a single modified and instrumented precision C(T) specimen. The C(T) thus calibrated could then be used for subsequent fixture measurements.

As an outgrowth of this work, a similar technique has been used successfully in correcting a crack front skewing problem in wedge-loaded crack arrest tests.

While the device was developed as a tool to aid in mechanical test setup, it may also be useful in verification of in-service loads on actual pinned components.

Conclusions

A simple instrumented device has been developed as a means of quantifying specimen bending loads of pinned clevis-type fixtures. The device can measure fixture machining contributions to alignment as well as axial eccentricity influences.

Acknowledgments

I wish to thank my colleagues J. J. Pajot for creating the finite-element model, W. W. Wilkening for performing the finite-element analyses, and B. M. Furbeck for his strain-gaging skills and general assistance.

Reference

[*1*] deLorenzi, H. G., "On the Energy Release Rate and the J-Integral for 3-D Crack Configurations," *International Journal of Fracture*, Vol. 19, 1982, pp. 183–193.

Willis D. Bowman[1]

End Constraint and Alignment Effects in Three- and Four-Point Reverse Bending Tests

REFERENCE: Bowman, W. D., "**End Constraint and Alignment Effects in Three- and Four-Point Reverse Bending Tests,**" *Factors That Affect the Precision of Mechanical Tests, ASTM STP 1025,* R. Papirno and H. C. Weiss, Eds., American Society for Testing and Materials, Philadelphia, 1989, pp. 174–184.

ABSTRACT: A specimen loaded in bending when rigidly clamped on the ends by specimen grips will undergo an axial force due to the conflict between the grips' resistance to move and the specimen ends' intention to move inwards. By mounting the grips on flexures and allowing them to move with the specimen ends while still applying a bending moment, pure bending on the specimen will result. A constant moment along the length of the specimen can be obtained by applying equal and opposite force couples to the ends of the specimen via the grips.

This paper examines the effects of specimen end conditions and alignment on test results obtained from three- and four-point reversing bend tests. Various end conditions can exist, such as fixed-fixed, fixed-free, free-free, and a special case in which the end condition is a combination of fixed and free. Each of these end conditions imposed on a specimen in bending is evaluated as to its contribution to axial forces and its ability to provide a true bending moment on the specimen. The effects of bending out of plane with respect to the specimen axis are also discussed.

This paper also describes an apparatus for testing specimens in three- and four-point reverse bending in which the axial force in the specimen is reduced.

KEY WORDS: three- and four-point bend test, reverse bend test, specimen alignment, flexures, boundary conditions, end constraints

To perform a true bending test on a specimen, any axial force or effects from a twist in the specimen must be eliminated. Historically, a specimen has had a bending moment imposed on it by applying a force (via a smooth roller) through its center (three-point test) or two forces at a certain span (four-point test) with two supporting rollers at a span outside of these. This bending moment method is acceptable for nonreversing loads (that is, for forces which may be cyclic yet do not reverse or pass through zero), for the specimen is allowed to rotate around the roller and needs to be constrained only in the direction opposite to the force. The specimen will always contact the roller tangentially, producing a contact force only; thus, no axial force is produced in the specimen.

The situation becomes more complex when a specimen undergoes a reversing bend test, for example, a cyclic fatigue fully reversing test. Now that the force couple (the bending moment) is reversing, the specimen must be constrained in two directions as opposed to one in the nonreversing test, yet the constraint must not create any axial force in order to obtain true bending moment test results.

Another consideration of performing the pure bending moment test is that the vector

---

[1] Product engineer, MTS Systems Corp., Minneapolis, MN 55424.

(or thrust) of the applied moment must be parallel with the centroidal plane of the specimen. Any misalignment will cause a twist in the specimen, leading to unwanted shear stresses.

This paper will describe the details of three- and four-point reverse bend fixtures and those attributes which affect the precision of a pure bending test.

Theory of Bending

A body is said to be in pure bending when the neutral axis (a contour line of zero stress) and the centroidal axis (a contour line which passes through the body's moment of inertia) of the body correspond, that is, when there is no stress at the centroid as shown in Fig. 1. The only way of producing such a bending stress field is to completely eliminate any axial force.

Figure 2 shows that the neutral axis of a specimen under a combination of bending and axial force shifts away from the centroidal axis, causing a higher stress on the bottom fiber than the top.

It is important to note that in practice almost all members in loaded structures have components of axial, bending, and shear stresses. With anisotropic materials and substrates such as composite materials and ceramics, it is advantageous to isolate characteristics corresponding to different loading schemes (tension, compression, bending, shear, torsion, etc.). The fixture described below minimizes the axial force so that a specimen can be evaluated during a pure bending test.

There are several ways to apply a moment to a specimen. One way is to apply a force at the midpoint of the specimen's length, creating a triangular moment distribution as depicted in Fig. 3a. An alternate method is to apply a moment (force couple) at the ends of the specimen, which will experience a rectangular moment distribution as shown in Fig. 3b. Figure 3c is a representation of the moments from Fig. 3b broken into two force couples.

Applying a moment at the ends of the specimens often produces better results than the other two methods for two reasons. First, an even moment distribution (constant along the specimen) allows for a variety of specimen geometries. A symmetrical part will usually fail at the point of highest bending moment concentration; hence, that part loaded under a constant bending moment distribution can be fabricated with considerations independent of the moment distribution since it never varies. This can become an imporant consideration when costly prototype specimens are tested. Second, if a specimen is very soft

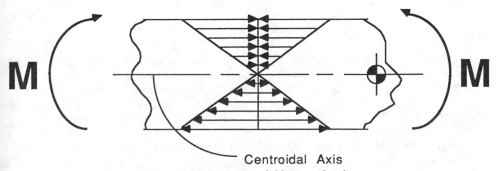

Centroidal Axis

FIG. 1—*Specimen's stress field in pure bending.*

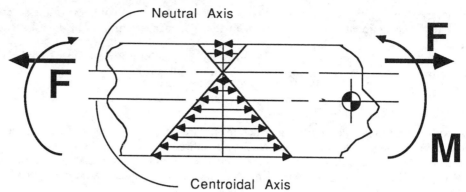

FIG. 2—*Specimen's stress field in bending and axial load.*

or brittle, it can be damaged by hertzian stress when a transverse force is applied. A typical method of applying that force is pushing on the specimen with smooth rollers. Because of the detrimental effects caused by the three-point bend test, it is recommended to strain gage a dummy specimen to determine the amount of shear stress caused by the center roller. If the shear stress is too high, then the four-point bend test should be used. Further details about specimen preparation are presented in the fixture discussion.

Boundary Conditions

The choice of boundary conditions (end constraints) has considerable consequences on the outcome of the bending test. As stated, the presence of an axial force applied to the specimen is detrimental to the bend test, so the boundary conditions must be chosen to eliminate (in theory) or at least minimize (in practice) this axial force.

The following is a discussion of various boundary conditions set on the specimen and the benefits and detriments of each. To repeat, these cases are for fully reversing forces.

FIXED-FIXED

Although easy to incorporate within a fixture, the case shown in Fig. 4a tends to impose a substantial axial force in the specimen. The conflict between the tendency of the ends of the specimen to move inwards as a radius forms, and the end constraints resistance to move forms axial forces.

FREE-FREE

The case in Fig. 4b can be disregarded, as there are no means of holding the specimen down during reversing forces; however, no axial force is developed as the ends of the specimen are allowed to travel horizontally when the specimen follows a concave shape.

FIXED-FREE

This condition suffers from the same problem as in the FREE-FREE case; the right end of Fig. 4c would lift up in a fully reversing cyclic test.

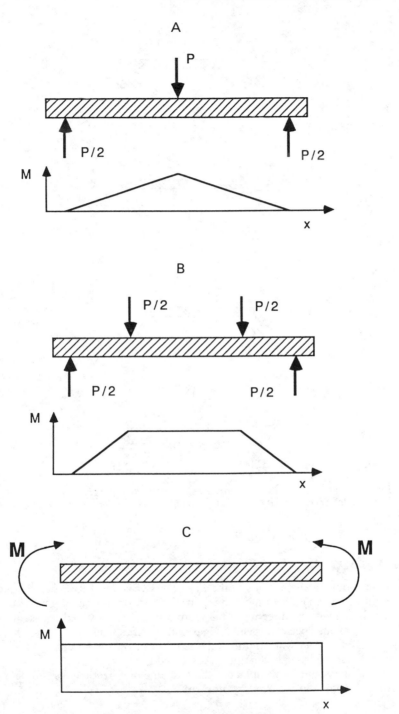

FIG. 3—(a) *Moment diagram for specimen in three-point test.* (b) *Moment diagram for specimen in four-point test.* (c) *Moment diagram for specimen with constant moment applied.*

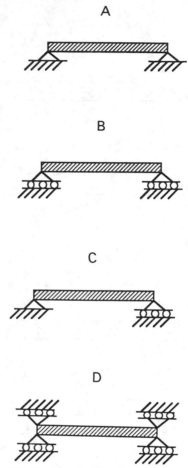

FIG. 4—(a) *Specimen with fixed-fixed end constraints.* (b) *Specimen with free-free end constraints.* (c) *Specimen with fixed-free end constraints.* (d) *Specimen with fixed-fixed end constraints.*

FRIXED—FRIXED[2]

This case combines the benefits of the FIXED-FIXED and FREE-FREE conditions by allowing the specimen ends to move freely in the horizontal direction while constraining movement in the vertical direction. This setup also allows torsional release to permit the ends to rotate as the specimen bends. Figure 4d illustrates this situation. This release in the horizontal direction will minimize the axial force on the specimen. The manner in which this is accomplished in a fixture is discussed later.

As the boundary conditions influence the outcome of a bending test, so does the alignment of the specimen to the bending moment.

---

[2] This is a pseudonym for a combination of FIXED + FREE = FRIXED.

Specimen Alignment

In order to obtain a pure bending moment in a specimen, no axial, shear, or torsional forces may be present. The preceding section offered a scheme to reduce or eliminate the axial forces. The shear and torsional stress may be reduced or eliminated by aligning the specimen to the bending moment inducing fixture. The specimen axis must be perpendicular to the vector (thrust) of the applied bending moment, and these bending moment vectors, one on each end of the specimen, must be coplanar to the specimen's neutral axis as shown in Fig. 5a. Figure 5b illustrates nonparallel bending moments which are not perpendicular to the neutral axis. This scenario will cause another couple to form whose vector is along the specimen axis. The ultimate situation is to create a force array where only bending stresses are produced and no secondary forces are developed due to misalignment.

Figure 6 demonstrates the case in which the specimen support points are not coplanar (in this instance, the bending moment vectors pass through the specimen). This offset,

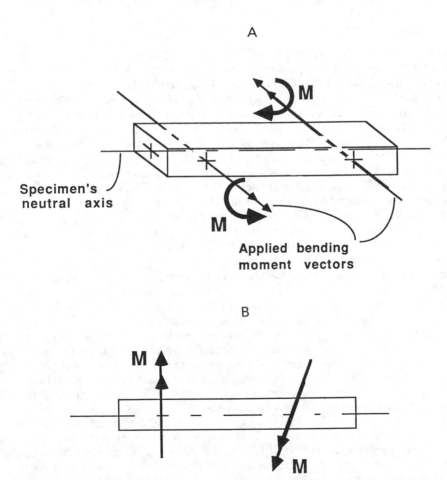

FIG. 5—(a) *Neutral axis and moment vectors of a specimen in pure bending.* (b) *Nonparallel bending moments applied to a specimen will cause a twist.*

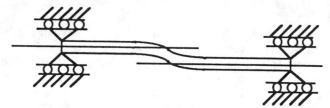

FIG. 6—*Noncoplanar specimen clamping surfaces will create shear in the specimen.*

regardless of the applied force, will cause a shear stress in the specimen. When shear, torsional, and axial forces are reduced or eliminated by use of correct boundary conditions and alignment of the specimen to the bending moment fixture, then pure bending will result. The next section details how these facts are incorporated into a reversing bend fixture.

The Reversing Bend Fixture

As seen in Fig. 7, the method of controlling the axial force is accomplished by flexures and tapered roller bearings. By referring back to Fig. 4d, the concept of FRIXED (FREE and FIXED) is translated into hardware; the flexures allow the specimen's ends to move horizontally yet restrain them from displacing vertically, while the bearings allow a torsional release at the ends as the specimen forms a radius. The design of the flexures is very critical to the success of the bend fixture. They may be thought of as thin cantilever beams with an axial and transverse force applied. They must be thin to be as flexible as possible in the axial direction, though thick enough to withstand the axial and bending fatigue stresses and buckling stresses. The actual applied bending moment is formed by the force couple transmitted through the upper and lower flexures. The applied force is generated by an actuator connected to the lower flexure plate.

The applied moment is transferred from the flexures to specimen holding blocks via tapered roller bearings. It is imperative that the bearings be backlash-free to eliminate any discontinuity in the force as it passes through zero. This is accomplished by preloading the cones against the cups with shims. The bearing shaft, as well, is preloaded against the specimen reaction block to negate any backlash. As the energy of the induced moment is transmitted to the specimen through the bearings, they must have low friction to minimize the transmittal of extraneous noise into the load train.

It is important to note that because of the nature of this test and fixture, there will be an axial force (though small relative to the fixture input force) in the specimen caused primarily by the stiffness of the flexures and secondarily by friction in the bearings. With any test fixture it is important to quantity how much the fixture is skewing the test results, therefore strain gaging a dummy specimen is recommended to test for axial force. The author has found the axial force approximately 0.5 and 1% of the fixture input force in the three- and four-point configurations, respectively.

The specimen must be clamped rigidly to the test fixture, for slippage will result in erroneous data in the form of high strain, low forces, and noise, or even damage to the specimen. If the specimen can bear it, it is beneficial to clamp it between textured surfaces (serrations, sandblasting, or weld texturing) for higher friction. The amount of clamp force on the specimen is determined by the coefficient of friction between specimen and clamping plates, the force on the specimen, and the amount the specimen will yield during clamping (causing an axial force due to Poisson's effect). As stated earlier, the four-point

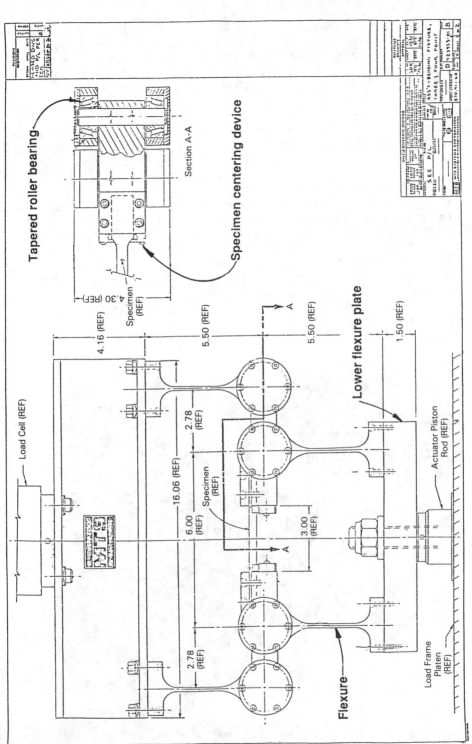

FIG. 7—*Four-point reversing bend fixture layout.*

bend test was preferred because of a constant bending moment distribution allowing a specimen geometry independent of that distribution. Though easy to fabricate, a specimen with a constant square or rectangular cross section may not prove to be the best geometry as end effects from grips (that is, high compressive clamp force) will cause failure in the grips rather than in the center. The ideal specimen configuration is the dog-bone shape; however, this may not be feasible with hard-to-manufacture materials as composites and ceramics.

This bending fixture has been designed with a specimen-centering device that centers the specimen to the midline of the fixture, which in turn fixes the centroidal axis perpendicular to the bearing axis, thus eliminating specimen twist. Figure 8 shows a rack and pinion assembly which centers the specimen to the fixture.

Figure 9 is a photo of a completed bend fixture in the four-point configuration.

In the three-point configuration, the force application rollers are made of smooth finished, hardened steel. As mentioned previously, rollers may cause damage to a specimen through hertzian stresses. In this instance, this is difficult to remedy, as the particular test for which the fixture was designed requires a force in the center. The roller's axes, like the bearings, are perpendicular to the specimen's centroidal axis. Figure 10 illustrates the bend fixture in a three-point configuration. Note that the outside flexures are disconnected from the top flexure plate and that a center roller support is added.

All connecting plates and parts through the load train are machined flat and parallel to minimize angularity and concentricity errors.

The operation of the fixture is straightforward, yet care must be taken if brittle or low-

FIG. 8—*Rack and pinion specimen-centering device assembly.*

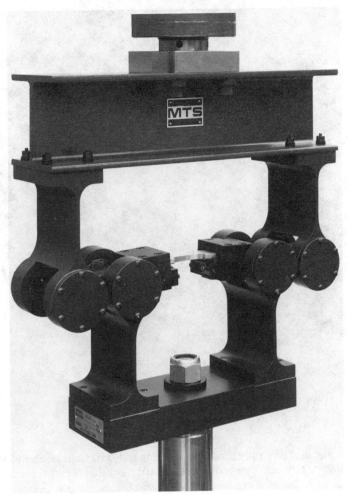

FIG. 9—*Reversing bend fixture in a four-point configuration.*

strength materials are being tested. In the three-point configuration, the specimen is clamped in the fixture and the upper center loading roller is brought into contact with the top of the specimen. The lower center loading roller is held on the underside of the specimen while the two retaining bolts are tightened. The actuator must not drift when loading the specimen or damage may result.

In the four-point configuration, the actuator must be positioned so that the specimen-holding blocks are coplanar. This is to eliminate any premature bending or failure of the specimen. The specimen is loaded and clamped down securely. Again, the actuator should not drift while loading.

Historical

Three of these bend fixtures have been built to date, although no data are available to determine their performance.

FIG. 10—*Reversing bend fixture in a three-point configuration.*

Conclusion

In order to perform a true fully reversing bending test, axial, shear, and torsional forces must be eliminated. The end constraints (boundary conditions) and specimen alignment play a large role in reducing these forces. It is found that a boundary condition, which is released in a horizontal direction, fixed in the vertical direction, and released torsionally in the direction of the others will eliminate or reduce axial force. This combination is conveniently produced by a bearing mounted in a flexure. The shear and torsional forces are lessened by aligning the specimen with the fixture and by providing coplanar specimen-mounting surfaces.

The induced bending moment is created by a couple at the end of the specimen yielding a uniform bending moment field along the specimen length. This loading scheme (four point) has proven better than loading the specimen in bending with a center force (three point) for two reasons: The bending moment field is not constant, and the center force may damage the specimen.

Acknowledgments

The concept for this three- and four-point bend fixture was originally developed by the Ford Motor Co. Several versions similar to Ford's have been built by MTS Systems Corp.

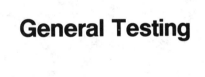

General Testing

Thomas G. F. Gray[1] and James Sharp[2]

Influence of Machine Type and Strain Rate Interaction in Tension Testing

REFERENCE: Gray, T. G. F. and Sharp, J., **"Influence of Machine Type and Strain Rate Interaction in Tension Testing,"** *Factors That Affect the Precision of Mechanical Tests, ASTM STP 1025,* R. Papirno and H. C. Weiss, Eds., American Society for Testing and Materials, Philadelphia, 1989, pp. 187–205.

ABSTRACT: The measured yield strength and strain rate sensitivity of a sample steel characterized by a yield point instability were found to be much lower when servocontrolled machines were used when compared with open-loop machines. Several machine types and control systems were used to test the sample steel at different rates, and measurements of gauge length plastic strain rate were made. The data are used to evaluate the machine "hardness" strategy for calculating or predicting the plastic strain rate developed in a machine which has no direct strain rate control. The background to current procedures embodied in current British standards for tension testing is discussed, and tentative proposals are made to classify machine types and set appropriate testing rates so that the results of tension tests using different systems may be reconciled.

KEY WORDS: upper and lower yield strength, ultimate tensile strength, yield drop, strain rate sensitivity, servocontrol, hardness ratio in tension testing, testing machine compliance effects, British Standards 18 and 3688

Nomenclature

A Specimen cross-sectional area
E Young's modulus
K Machine compliance
l Specimen length
l_1 Total parallel length
l_2 Extensometer gauge length
H Measured hardness ratio $(\dot{u}_{pl}/\dot{u}_{el})$
H_2 Calculated hardness ratio
$\dot{u}_{el}$ Elastic extension rate in gauge length
$\dot{u}_{pl}$ Plastic extension rate in gauge length

Introduction

The aim of the project described in this paper was to find out how much variation in the tensile properties of a given steel could be attributed to the use of different testing machines and procedures. Current procedures in U.K. standards are based on test work of 20 years ago, and significant developments have taken place in testing machines since that

---

[1] Reader, Division of Mechanics of Materials, University of Strathclyde, Glasgow, Scotland.
[2] Research assistant, Materials Testing Laboratories, University of Strathclyde, Glasgow, Scotland.

time. A particular objective, therefore, was to generate information related to new machine technology which would assist the process of standards revision. Many machines of an older generation are still in use in test houses, however, and the project was also designed to yield information relevant to older machines.

The accurate determination of yield is of considerable commercial importance aside from its technical significance in design and service performance. Many grades of structural metals, steels in particular, are specified in terms of a minimum upper or lower yield strength at a given temperature. In any batch there is an inevitable variation, and any variation due to testing procedures can lead to rejection of material or downgrading to a lower-priced class. Producers are well aware of these problems and have learned to use any looseness in the testing standards to avoid such penalties. There is also an international dimension to the material release question, as producers in different countries are in competition. Although the testing standards are very similar throughout the world due to the efforts of ISO, again any looseness in specification results in the divergence of procedures, resulting in unfair comparisons between materials produced in different countries.

For these reasons, the U.K. Department of Trade and Industry was persuaded by the British Standards Institution to fund the preliminary project described here, and further industrial support in kind has been received from other sources.

Background

It has long been understood that the strength properties exhibited by materials may depend as much on the testing system as on the materials themselves. David Kirkaldy [1] was one of the early pioneers of systematic testing and identified the role of strain rate over 100 years ago, despite using equipment which would be considered primitive by today's standards. The effect of variable *loading* rate was demonstrated by Professor Archibald Barr in an illustrated lecture given in 1908. He used an ingenious dead-weight autographic wire testing machine and concluded that "the apparent mechanical properties of materials depend greatly on their prior treatment and manner of testing. It is therefore rather ridiculous to quote yield strength to several decimal places, . . ." [2]. Writing in *The Engineer* in 1934, J. L. M. Morrison argued, "If, then, it is agreed that the yield stress of the materials is of paramount importance it is obvious that the definition of the yield must be standardized and the method of measuring it above suspicion." References [3–7] represent a small selection of more recent work which examines the effect of testing machine characteristics on apparent properties.

The present U.K. standard for tension testing ("British Standard Method for Tensile Testing of Metals," BS18: 1987) is based largely on development work carried out in the 1960s [8,9] when open-loop hydraulic machines were the norm in test houses. Control of crosshead velocity of such machines was of variable quality; better examples were fitted with pressure-compensated valving to maintain constant oil-flow rate as load increased. The conclusion of this research was that the test system could be represented as an assembly of linear springs up to the yield point, with the addition of a dashpot-like element thereafter (see Fig. 1). This representation led to a simple relation between plastic and elastic extension rates in the yielding section of the specimen

$$\dot{u}_{pl}/\dot{u}_{el} = 1 + KAE/l \tag{1}$$

(The plastic/elastic rate ratio is sometimes called the "hardness" ratio.) A similar formulation is contained in the German DIN standard, which refers to Ref *10*.

The British standard based on these findings was first published in 1971. It restricted

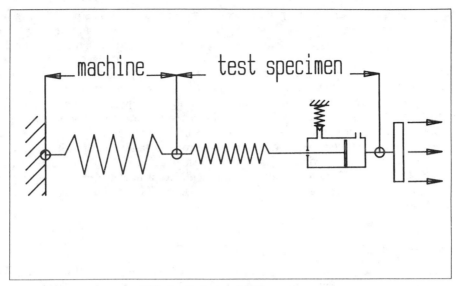

FIG. 1—*Spring model of test system.*

plastic strain rates useable in the determination of lower yield strength (LYS) to the range between 0.015 and 0.15/min as the mean effect on LYS over this range had been found to be of the order of only 14 N/mm^2. The question of how tests could be controlled to stay within this range was addressed by issuing a related standard ("British Standard Method for the Determination of K Values of a Tensile Testing System," BS 4759: 1971), which gave a procedure for measuring the "compliance" of a given testing machine system. Essentially, the user was asked to carry out a sample test, as near to the proposed conditions of use as possible, and measure specimen elastic and plastic strain rates. The resulting compliance, determined through inversion of Eq 1, could then be used in other tests to set an elastic strain rate (or stress rate) which would provide plastic strain rates in the required range. In the DIN standard, the testing machine manufacturer is expected to supply the "hardness" ratio value.

A separate standard for elevated temperature ("Methods for Mechanical Testing of Metals at Elevated Temperatures," BS 3688: 1971) restricted strain rates to a narrower range, more than an order of magnitude slower (0.001 to 0.003/min). In this case the use of specimen-mounted extensometers is mandatory, and it is thought that the slow rates specified were chosen to facilitate extensometer reading. BS18 also calls for similarly slow strain rates (0.001 to 0.005/min) when testing aerospace materials ("Category 2 materials") for proof strength. Such materials have not been addressed so far in the present project, but it is of interest to explore the capability for operating machines at these levels.

The present position in the standard concerning the use of extensometry in room temperature tests is rather subtly stated. For all *proof* strength determinations, extensometry to specified grading must be used. However, upper and lower yield strengths may be determined "visually" (a flexible description) "or by a force/extension diagram." In the latter case, it is permissible to record displacement from the crosshead motion. In practice, "visual determination" may be from hesitation of the load pointer in older machines or from rate of change of the digital display on newer machines; but the standard very wisely does not elaborate.

Substantial changes have taken place in testing machine technology since 1971. In particular, servocontrol has provided more consistent control of crosshead rate, or even of specimen strain rate if a feed-back extensometer is fitted. Modern machines also tend to be more powerful and more complex from the verification point of view. However, extensometer control is still not popular for routine testing, and it appears that use of the compliance calibration procedure is more honored in the breach than in the observance. The present position is that the compliance standard has been proposed for withdrawal, and the most recent revision of BS 18 makes no reference to any other method for strain rate control in open-loop or non-extensometer-feedback arrangements.

These circumstances, together with pressure in European standards committees to revise testing rates upwards, pointed up the need for a fresh examination of the basis for tension testing procedures in the light of modern equipment. Accordingly, the U.K. Department of Trade and Industry supported a preliminary project to this end under the guidance of the relevant BSI committee.

Project Test Machines

Tests were carried out on a common material sample using six different machine types. Results from four of these machines have been analyzed for this paper. In two cases where servocontrolled machines were used, these were applied in alternative control modes, namely, ram-displacement feedback and specimen-mounted-extensometer feedback. The various types of machines and control systems are designated in this paper as follows:

Open-loop mechanical (screw-driven crosshead)		OLM
Open-loop hydraulic		OLH1
Closed-loop hydraulic—Ram displacement feedback		CLH-RD
	Extensometer feedback	CLH-E
Closed-loop hydraulic	Ram-displacement feedback	CLH-CC1-RD
with computer	Extensometer feedback	CLH-CC1-E
control		

Thus, six different combinations of frame and control system are treated in the analysis.

The two open-loop machines have mechanical or electromechanical load indicating systems, whereas the servomachines have stiffer electronic load cells. The open-loop hydraulic machine is of pre-1970 design and does not feature the more accurate oil-flow regulation circuits which have since been incorporated in a virtually identical frame. All tests were conducted at constant crosshead rate (or at least with controls "untouched" during the test). In the case of the computer-controlled machine, this meant that the software, which varies the speed depending on the stage of the test, was inhibited.

Test Material

The specimens were taken from a single 30-mm-thick plate of BS 4360 Grade 43 steel with a test certificate yield strength of 314 N/mm^2. A typical chemical analysis is given in Table 1. This material exhibits upper and lower yield points and a fairly flat postyield characteristic. The material was machined to give profiled cylindrical specimens of BS 18 preferred dimensions, varying in four steps of cross-sectional area between 50 and 200 mm^2 (corresponding to 8 to 16-mm diameter approximately).

TABLE 1—*Material specification and analysis.*

Material Specification	Chemical Analysis, %										
	C	Si	Mn	P	S	Cr	Mo	Ni	Al	Cu	V
BS 4360 1979 Grade 43E	0.143	0.226	1.265	0.011	0.018	0.017	.004	.017	.05	.01	.001
Steel plate, BOS Process, normalized at 910°C—45 mins											

Test Instrumentation and Procedures

In all cases, regardless of the mode of control, a 50-mm gauge length extensometer was fitted to give instantaneous output of specimen strain. Signals proportional to load were derived from the integral load cell in the case of the servomachines or from a specially fitted transducer in the mechanical load measuring system of the two open-loop machines. These signals were then applied to two A3-size XY plotters to give simultaneous load/extension and extension/time graphs (see Fig. 2). This instrumentation was almost essential for setting up the required elastic loading rates on the open-loop machines, and as the test program developed, full elastic/plastic strain-time graphs were generated. This allowed direct measurement of "hardness" ratio in terms of Eq 1.

The test procedures were designed to satisfy two objectives: determining the influence of strain rate on properties and discovering the effect of letting the machine run in the manner prescribed by standards. A program of strain rates was set to cover the ranges prescribed by BS 18 and BS 3688, plus a decade or so above and below.

In the case of the extensometer-controlled tests, the desired strain rates were set directly using the machine controls, and the strain rate records showed that these rates were obtained very consistently. It is not clear what practice is typically adopted by test houses

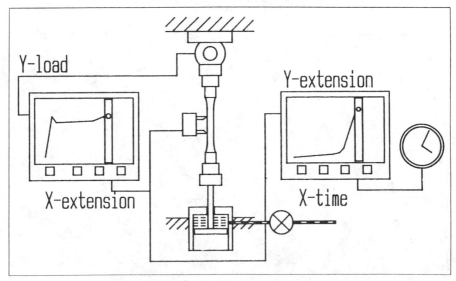

FIG. 2—*Test instrumentation.*

in the case of tests conducted without extensometer feedback, but the likelihood is that if any parameter is set and checked, it is probably elastic loading or stressing rate. Hence elastic strain rates in the regime before upper yield were set as targets in the present test series. The resulting plastic strain rates were then simply measured and accepted as a variable. The combination of instrumentation and procedure described above was such that accuracy of control and measurement of strain rate were better than is normally obtained in routine testing.

Results

Lower Yield Strength

Effect of Machine Type. The *extensometer-controlled* test results can be taken as a base line for the other tests. Figure 3 shows a typical test result at the upper end of the BS 18 strain rate range. All extensometer-controlled results are given in Fig. 4 together with upper and lower 95% confidence levels on a log/log linear fit. (A data weighting scheme was used prior to the fitting process in accordance with Ref *11*—see Appendix.) The ranges of plastic strain rate permitted by BS 18 and BS 3688 are shown in the figure. In this case, there is no doubt that the measured rates were maintained through the elastic regime during the yield drop and on to the lower yield plateau.

The apparent sensitivity to strain rate is about 40% less than has been shown in previous studies on this type of material [9], and in this case the scatter band is reasonably tight (± 12 N/mm^2). It would be unreasonable to expect much less scatter than this, given that there are normally point-to-point property variations in a commercial steel plate and that material as characterized by yield instability is always susceptible to variation.

In the case of the *non-extensometer-controlled* tests, it is not so obvious how these should be plotted. Following the idea that initial stress rate is the parameter most likely to be controlled in a practical test, measured initial elastic strain rate is used in Figs. 5–7 where

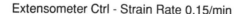

Extensometer Ctrl - Strain Rate 0.15/min

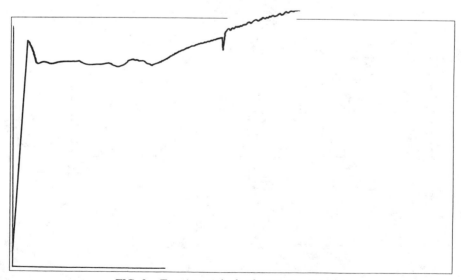

FIG. 3—*Test material—load extension autoplot.*

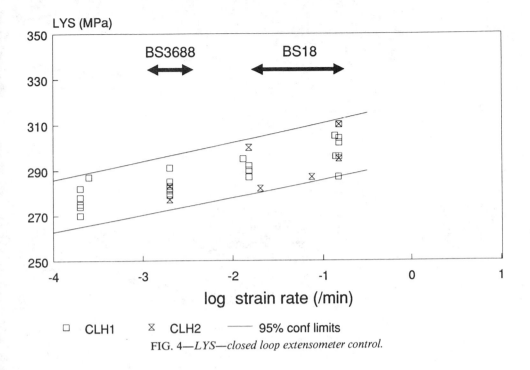

FIG. 4—*LYS—closed loop extensometer control.*

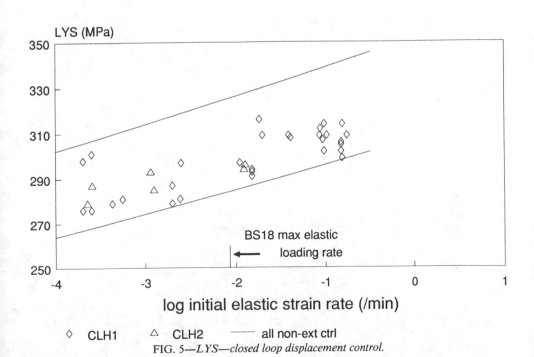

FIG. 5—*LYS—closed loop displacement control.*

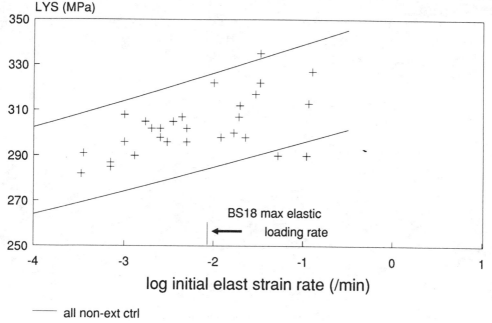

FIG. 6—*LYS—open loop mechanical.*

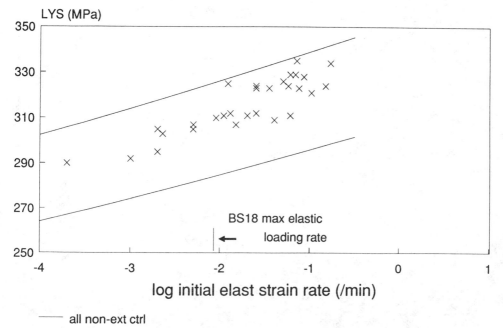

FIG. 7—*LYS—open loop hydraulic.*

the performance of different machine types is shown against a scatter band for all non-extensometer-controlled tests. The apparent strain rate sensitivity and the scatter are quite large (± 21 N/mm^2), partly because the "hardness" ratio for individual tests varies markedly, and hence a given initial elastic strain rate produces a wide variety of plastic strain rates following yield. The strain rate equivalent of the BS 18 guideline for stress rate is also shown. It is clear from these figures that initial elastic strain rate or stressing rate is an unsatisfactory control parameter, especially when it is noted that the data shown are in terms of *measured* elastic strain rates, as distinct from rates *set* on the machine. In practice the elastic strain rate is often set on the basis of a simple crosshead rate calculation.

Note from Fig. 5 that the servocontrolled (ram-displacement) results fall to the lower half of the scatter band and themselves show less scatter. The open-loop data in Figs. 6 and 7 are mainly responsible for the steeper strain rate sensitivity and wider scatter of the overall band. In fact, it proved very difficult to control the hydraulic machine down to a level which brought the plastic strain rate within the BS 18 band. The strain rate sensitivity is about the same as shown in previous studies which were quoted in terms of plastic strain rate. Examination of individual results shows that none of the servocontrolled tests conducted at initial *elastic* rates below the BS 18 guideline exceeds the plastic strain rate upper limit of 0.15/min. However, several of the open-loop results break the limit, although the initial loading rate was below the prescribed level. Detailed examination of the results showed that if the "hardness" ratio was above a value of about 20, the recommended elastic stress rate limit was too high.

Non-extensometer-controlled results where steady postyield plastic strain measurements were made are plotted in Fig. 8 in terms of the extensometer strain rate. This removes much of the scatter due to variation in "hardness" ratio, but comparison with the extensometer-controlled scatter band shows that the strain rate sensitivity is still 36% greater. However, the indicated yield strength is effectively the same for all control system types at

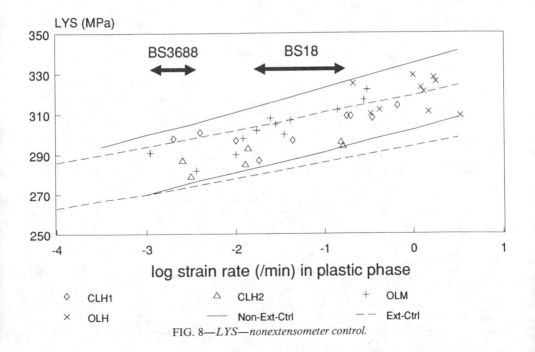

FIG. 8—*LYS—nonextensometer control.*

strain rates a little below the BS 3688 minimum. It does not initially seem reasonable that there should be a difference between the two systems when measured plastic strain rate is used to plot the results, but explanations will emerge in a later section on upper yield strength.

Given that initial elastic strain rate is seen to be an inadequate control parameter, it is of interest to establish the extent to which the original strategy of using a calibrated "hardness" to predict plastic strain rate actually works. The apparent compliance can be back-calculated from the measured elastic and plastic strain rates via Eq 1, but a variation of that equation is necessary to allow for the parallel length of specimen which lies outside the extensometer gauge length in the larger specimens. In that case "hardness" ratio H is given by

$$H_2 = \dot{u}_{pl}/\dot{u}_{el} = (KAE + l_1)/l_2 \qquad (2)$$

When the data were analyzed, it was clear that the apparent compliance was far from constant on a given machine. The variations are shown in Table 2. There are insufficient data to discern all the trends clearly. However, part of the variation relates to the fact that the elastic stiffness of the "frame" is usually nonlinear, being more flexible at lower loads (smaller specimen areas).

This was recognized in the original compliance measurement standard, which suggested that the machine should be calibrated at the load level to be used in testing. Another effect is more subtle, being that the "frames" appear to be stiffer at higher testing rates, or more correctly, the "hardness" ratio is less at high strain rate. The problem is that the slower the strain rate, the more plastic strain develops in the specimen relative to the elastic strain. Thus "hardness" and apparent compliance are essentially time-dependent parameters, and the simple assumption of a spring model is not valid.

Ignoring all these difficulties, the *average* compliance for each machine, over all tests shown in Fig. 8, was used to predict the theoretical plastic strain rates from the measured elastic rates. Comparison of the predicted rates with the measured rates is shown in Fig. 9. The rate effect on apparent compliance is seen in the deviation from 1/1 correlation, and there are some points which lie more than a half decade away from the mean. Notice, however, that the closed-loop servoresults display high correlation.

The scatter band for all non-extensometer-controlled tests was then calculated on the basis of predicted plastic strain and is shown in Fig. 10. (The upper limit of the "predicted" scatter band coincides with the "actual.") Figures 9 and 10 suggest that the "hardness" approach is quite good, given that careful measurements of hardness ratio are made and a wider scatter band is accepted. However, the elastic rates used in the predictions were "as-measured," and it would be much more difficult to achieve the required plastic rates if one had to rely purely on setting the crosshead displacement rates.

TABLE 2—*Performance of machine frames.*

Machine Designation	Compliance, mm/MN			Measured "Hardness" Ratio	
	min	max	mean	min	max
CLH-CC1-RD	20	30	25	10	14
CLH-RD	2.6	19	8.4	4	10
OLH1	37	325	107	13	84
OLM	0.36	114	21	2	24

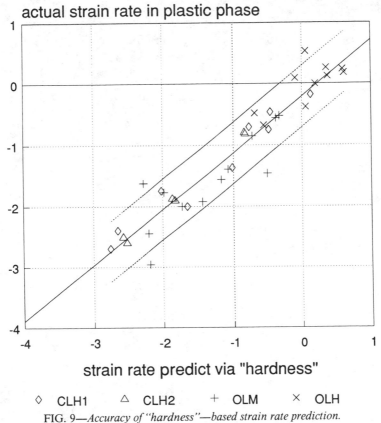

FIG. 9—*Accuracy of "hardness"—based strain rate prediction.*

Influence of Cross-Sectional Area

Clear trends of behavior with varying specimen cross-section are difficult to establish, as a full set of comparable data across the range of sizes and machine types was not available. However, the servocontrolled test results seemed to be relatively uninfluenced by specimen diameter. The open-loop test results, on the other hand, showed a clearer tendency for the smaller diameters to give higher levels of lower yield strength. Figure 11 shows the results from the two open-loop machines plotted in terms of the calculated plastic strain rates and classified by specimen cross-sectional area. The differences are more pronounced when plotted in terms of initial elastic strain rate.

Upper Yield Strength (UYS)—Effect of Machine Type

Accepted wisdom in testing standards is that upper yield strength is simply a function of initial elastic loading rate. BS 18 stipulates that if *only* UYS is being measured, a range of elastic loading between 9×10^{-4} and 9×10^{-3}/min for steel should be used. (The upper limit here also had an eye to dynamic effects in testing machines which had been found to cause upper-yield overshoot). Figure 12 compares the scatter bands for extensometer-controlled and other tests, plotted on the basis of elastic strain rate. At least part of the reason for the difference is that extensometer feedback checks the development of plastic strain

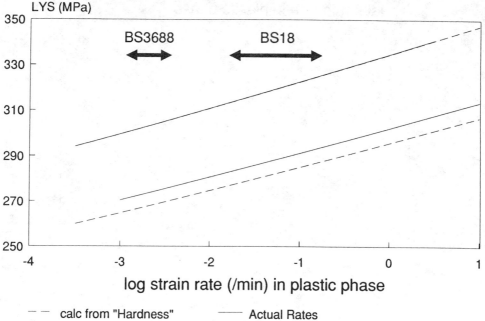

FIG. 10—*LYS—nonextensometer control.*

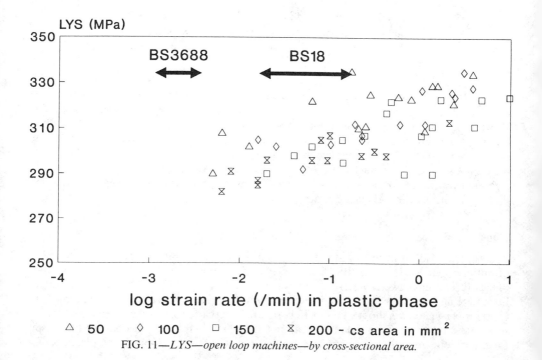

FIG. 11—*LYS—open loop machines—by cross-sectional area.*

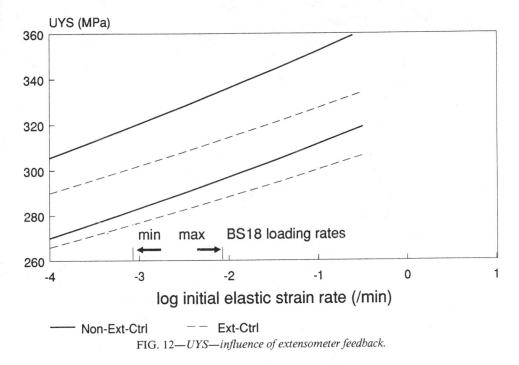

FIG. 12—*UYS—influence of extensometer feedback.*

following the onset of yield and gives the classical yield drop, whereas on the remaining machines the load continues to rise to a higher peak as the strain rate increases through yield. The non-extensometer-controlled results also show a much greater scatter. Once again, the servocontrolled results lie towards the bottom of the band for nonextensometer results.

The strain rate sensitivity is higher in all cases than for LYS, reaching a level close to that found in the original research [3,8,9].

The immediate post-yield phase seems to be fairly critical in influencing the value measured on the LYS plateau, and this gives a pointer to the differences in LYS between controlled and uncontrolled machines discussed with reference to Fig. 8. Figure 13 shows yield drop plotted against plastic strain rate for all servocontrolled tests, and a relatively smooth increase is seen with plastic strain rate. These results contrast with the behavior of open loop machines (see Fig. 14) where the effective yield strength tends to be maintained at higher strain rates (that is, little yield drop), resulting in a higher LYS plateau. What is slightly strange is that the elevated yield in these cases can be maintained well beyond the initial yield drop regime. As noted earlier from Fig. 8, the yield strength at a given measured plastic strain rate is higher for the nonextensometer controlled machines. It is believed, without detailed evidence at present, that the rapid response to overshoot of the extensometer-controlled systems and servosystems checks load increase and allows a greater proportion of the specimen gauge length to start yielding. As a result, a given measured plastic strain rate over the gauge length corresponds to a lower and more uniform local plastic strain rate than is the case for the uncontrolled tests. Hence the measured yield strength is lower for the extensometer-controlled examples, and even the displacement servocontrolled results are lower. It may even be that if the extension gauge could be precisely matched to the length which is actually straining plastically, even less strain rate sensitivity would be found for feedback-controlled test procedures.

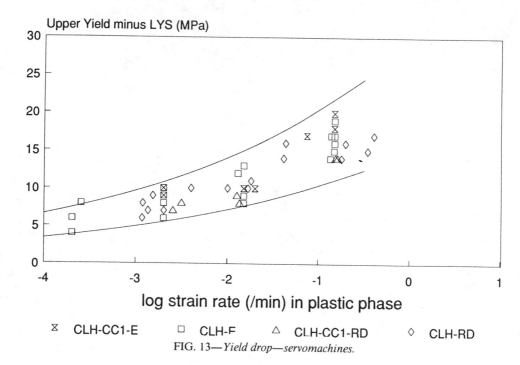

FIG. 13—*Yield drop—servomachines.*

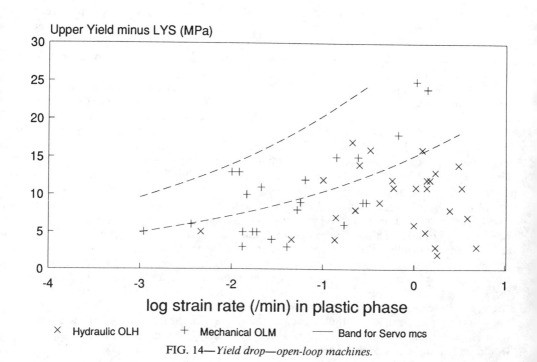

FIG. 14—*Yield drop—open-loop machines.*

Ultimate Tensile Strength (UTS)

The maximum tensile strength property is usually less critical from the design point of view, but it is nevertheless important to determine the variability due to testing method. In normal test-house practice, any extensometers used are usually removed before reaching the maximum load and the test is completed using some kind of crosshead-rate control. The control parameter in effect at this stage of the tests in the present project depended on the machine in use. In the case of the two extensometer-controlled machines, the crosshead-rate operative during controlled plastic straining in the early rising part of load/strain curve was maintained by switching to ram-displacement control at a suitable juncture. In the tests carried out under ram-displacement servocontrol or in open-loop, the crosshead rates were not deliberately altered. Hence the plastic strain rates would normally be somewhat less than the rates measured or calculated for the lower yield regime.

The maximum tensile strengths for tests where the plastic strain rate was confidently known are plotted in Fig. 15. It can be seen that the sensitivity to strain rate is fairly weak, and, although the results are scattered (± 13 N/mm^2), there is little systematic trend related to testing machine type. The maximum strain rate recommended in the current BS 18 is shown for comparison.

Discussion

Variation of Yield

The results of this survey do not give grounds for optimism concerning the precision of yield strength values measured in routine tension testing of non-work-hardening steels. The practices adopted in the project were probably more favorable to precision than nor-

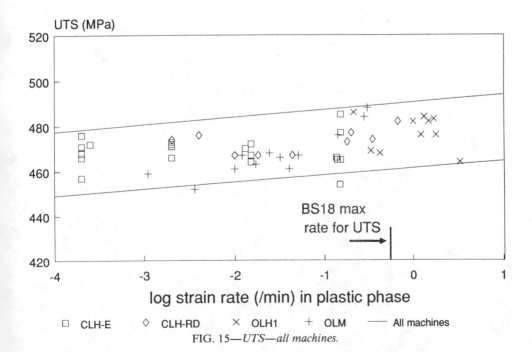

FIG. 15—*UTS—all machines.*

mal. For example, the variables introduced by testing unmachined stock were excluded, considerable care was taken to set the loading rates, and the additional nonstandard instrumentation provided a capability in this respect which is not normally available. Despite these improvements, the sample steel was found to have 95% confidence level variations as given in Table 3, and these, one suspects, are much larger than the users of such steels would find acceptable. Working strictly within BS 18, the sample steel could be shown to have a LYS anywhere between 279 and 341 N/mm². The picture is even more depressing when it is realized that the very conditions which improve precision, namely extensometer control, also deliver a lower level evaluation of LYS and UYS, in fact significantly lower at the maximum end of the BS 18 strain rate range (25 N/mm²). There is therefore little incentive given to producers to apply the best testing technology.

Control of Plastic Strain Rate

The experience of attempting to meet the aims of BS 18 in terms of controlling plastic strain rate also raises some doubts about the control precision achieved in routine testing. Where extensometer feedback is available, there should be little difficulty in implementing the standard rigorously, if not on every test, then at least for the purposes of calibration. The previous recommendations for indirect control of plastic strain rate, based on a measured "compliance," seem to have failed as judged by practical use. The results show, however, that the hardness ratio approach is perfectly viable for reasonably stiff machines with good control of crosshead rate. The penalty implied in using this prediction method is a greater scatter of yield values, amounting to another 6 N/mm² on the scatter associated with measured plastic strain values. This statement cannot be made as confidently for more flexible gripping systems or machines.

Standards Development

Comparison of the data from different machine types shows that when servocontrolled machines were used to test material with a yield instability, the results for yield strength tended to lie in a significantly lower band than the nonservo-controlled tests. It is something of a paradox that the measurement of yield in such materials is thought to require *less* control and instrumentation than the measurement of proof strength in mildly work-hardening materials. The opposite is in fact true. The confusion in approach has arisen simply because in normal testing practice the indicated load lingers about a particular level

TABLE 3—*Variation of designated properties.*

	BS 18 Strain Rate for LYS (/min)			
	Min (0.015), 95% Confidence		Max (0.15), 95% Confidence	
Property, N/mm²	Min	Max	Min	Max
LYS	279	313	287	341
UYS	290	339	302	356
UTS	at max rate of 0.5/min		462	487

TABLE 4—*Suggested classification of machines.*

Class	Control Mode	Instrumentation	System Compliance (Including Grips) Hardness Ratio
A	Extensometer feedback	Autographic plot	Not important
B	Crosshead/ram displacement feedback	Extensometer for calibration	<20
C	Open-loop-calibrated rate control	Extensometer for calibration	<20
D	As above	As above	<50

long enough for the observer to conclude that a specific defined value can be noted as a yield strength. In fact, this indicated level depends on the quality of strain rate control during the transition from elastic to plastic behavior.

If the aim in drawing up a standard procedure is to produce consistent results, then it is clear that the machine type forms a significant variable for such materials. The obvious conclusion is that a standard procedure will require some preliminary classification of machine types. Table 4 shows a first suggestion for such a classification. The target strain rate ranges would then depend on the class of machine in use. Figure 16 uses the present data to establish suitable limits which would introduce a measure of consistency between different machine types. Much more data is clearly required to underpin this proposal, but the present data serve to exemplify the procedure.

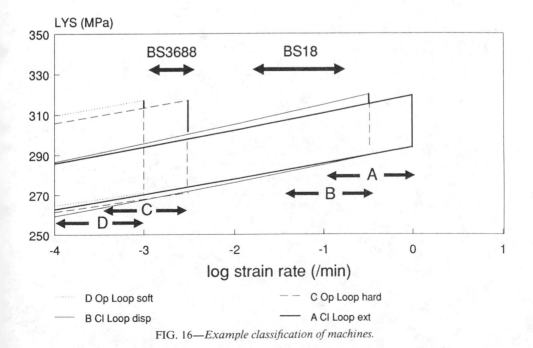

FIG. 16—*Example classification of machines.*

Conclusions

The measured yield strength and the strain rate sensitivity of a typical steel showing yield point instability are significantly less when servocontrolled testing machines are used. This is particularly the case when the feedback control signal is taken from a specimen-mounted extensometer. Tension testing standard procedures do not recognize this effect, and the use of servocontrolled machines therefore incurs a penalty on indicated yield strength relative to less well-controlled machines.

The use of frame compliance calibrations to predict the ratio of plastic to elastic strain rates in a test introduces fairly large errors in the control of strain rate, and the determination of yield strength is thereby made more inaccurate. This effect is also not recognized in standards.

Ultimate tensile strength is very much less sensitive to strain rate, and present procedures should give adequate precision.

There is a need to classify testing systems according to accuracy of control of strain rate so that measured yield values can be placed on a comparable footing. In cases where extensometer feedback control is not available, sufficient instrumentation should be provided and used to monitor the plastic strain rates actually achieved in a test.

Acknowledgments

We wish to thank Andrew Crockett for many hours of painstaking calibration and testing work and W. A. Shanks and A. A. F. McCombe, who contributed greatly to the project. The financial support of the U.K. Department of Trade and Industry is gratefully acknowledged together with industrial contributions by Avery-Denison, Instron, and British Steel Corp. The guidance and interest of British Standards Institution Committees ISM/NFM/4 and ISM/NFM/4/1 has also been invaluable.

APPENDIX

Weighted statistical fit to yield/strain rate data.

In conventional least-squares-fitting algorithms, each point is given the same weighting if the expected variance in each measurement of either variable is thought to be the same. This is a reasonable assumption for the data herein. If the data are transformed for plotting convenience by taking logarithms of one or both variables, a weighting of unity would then be incorrect and it can be shown that for log/log transformation the correct weighting factor is $w_i = y_i^2$, where y is the untransformed variable (assuming regression on x).

Thus the analysis in this paper is based on the following formulation:

$$SX = \Sigma(X_i w_i); \quad SY = \Sigma(Y_i w_i); \quad SXY = \Sigma(X_i Y_i w_i)$$
$$SX2 = \Sigma(X_i^2 w_i); \quad SY2 = \Sigma(Y_i^2 w_i); \quad W = \Sigma w_i$$
$$D = SX2 - (SX)^2/W$$

then

$$m = (SXY - SX*SY/W)/D$$
$$c = (SX2*SY - SX*SXY)/W/D$$

where

$Y = mX + c$ is the equation of the regression line,
y_i, x_i are untransformed coordinates, and
Y_i, X_i are transformed coordinates.

The Y offset for the confidence lines is given by

$$Y_{offset} = \pm t*[(SY2 - m*SXY - c*SY)/(W - 2)]^{1/2}*[1 + 1/W + (X - SX/W)^2/D]^{1/2}$$

where "t" is "student's" t.

References

[1] Kirkaldy, D., "Results of an Experimental Inquiry into the Tensile Strength and Other Properties of Various Kinds of Wrought Iron and Steel," Mitchell Library, Glasgow, 1862.

[2] Barr, A., "The Testing of Steel," *Journal of West of Scotland Iron and Steel Institute,* Vol. 17, No. 4, 1908.

[3] Lange, G., "Influence of the Hardness of the Testing Machine on the Results of the Tensile Test," *Archiv Eisenhuttenwes,* Vol. 43, No. 1, January 1972, pp. 67–75.

[4] Carmichael, A. J. and Betz, E., "Some Mechanical Characteristics of Testing Machines and Their Influence on the Response of Test Specimens," Institute of Engineers, Australia, *Mechanical, Chemical Engineering Transactions,* MC8(1), May 1972, pp. 100–103.

[5] Hass, T., "Determination of Material Behaviour in Tensile Tests Using Hydraulic Testing Machines," *Society Environmental Engineers Journal,* September 1976, p. 11.

[6] Mathonet, J. and Caubo, M., "The Tensile Testing Technique. Influence of Test Conditions on the Level of Measured Characteristics," Metallurgical Reports CRM26, March 1971, pp. 31–38.

[7] Tenaka and Ishikawa, H., "Effect of Rigidity of Testing Machines on the Behaviour of Tensile Deformation in Mild Steel," *Proceedings,* 19th Japan Congress of Materials Research, Iron and Steel Institute, Scarborough, England, 1976.

[8] Johnson, R. F. "The Measurement of Yield Stress" in ISI Publication 104, BISRA/ISI Conference, Iron and Steel Institute, Scarborough, England, 1967.

[9] Johnson, R. F. and Murray, J. D., "The Effect of Rate of Straining on the 0.2% Proof Stress and Lower Yield Stress of Steel," BISRA/1st conference, Eastbourne, England, Iron and Steel Institute, Scarborough, England, April 1966.

[10] Kravcenko, V., "The Behaviour of Steels Under Tensile Stress Below the Upper Yield Point," *Material prufung,* Vol. 12, No. 11, 1970, pp. 373–377.

[11] Guest, P. G., "Numerical Methods of Curve Fitting," Cambridge University Press, London, 1961.

John G. Lenard[1] and Achilles N. Karagiozis[2]

Accuracy of High-Temperature, Constant Rate of Strain Flow Curves

REFERENCE: Lenard, J. G. and Karagiozis, A. N., **"Accuracy of High-Temperature, Constant Rate of Strain Flow Curves,"** *Factors That Affect the Precision of Mechanical Tests, ASTM STP 1025,* R. Papirno and H. C. Weiss, Eds., American Society for Testing and Materials, Philadelphia, 1989, pp. 206–216.

ABSTRACT: The components that make up a mathematical model of hot strip rolling include the conditions of equilibrium, roll deformation, friction at the roll-strip interface, and material behavior. The consistency and accuracy of the predictive capability of the model will depend on the quality and rigor of the representations of these components. This paper deals with one of the least tractable parameters—material behavior.

The difficulties associated with the determination of constitutive behavior are discussed. It is suggested that a standard method of testing for high-temperature strength be established.

KEY WORDS: flow curves, high temperature, constant strain rate, accuracy

Engineers in the metalworking industry have, among others, the task of designing the equipment they use and of analyzing the behavior of the metal during the deformation process. These two endeavors are, of course, connected, the common link being an understanding of the mechanics of plastic flow of the metal as well as its resistance to deformation.

These, coupled with appropriate boundary and initial conditions, form a mathematical model of the process under consideration, the predictions of which are then used in the design phase. It is simple and straightforward to apply Newton's law of equilibrium to a slab of the metal under stress. If more refined distributions of the dependent variables and their rates are to be determined, use may be made of accurate and "user friendly" finite-element routines. Difficulties are encountered, however, when material behavior and the boundary conditions at the die/metal interfaces are considered. In this paper the difficulties encountered during the determination of the material's resistance to deformation are considered.

The quality of predictions of mathematical models of metal-forming processes depends, in a very significant manner, on the method of determination and representation of the material's resistance to deformation. The commonly used techniques of testing as well as the presentation of the resulting data were reviewed in detail recently by Alexander [1] and Lenard [2], and, in general, the conclusion emphasized was that the state of stress during the actual forming should be similar to that during measurements for strength.

The metals' resistance to deformation at high temperatures is affected by the interaction

[1] Professor, Department of Mechanical Engineering, University of New Brunswick, Fredericton, N.B., Canada E3B 5A3.

[2] Ph.D. candidate, Department of Mechanical Engineering, University of Waterloo, Waterloo, Ontario, Canada N2L 5G1.

of several metallurgical mechanisms, resulting in hardening and softening phenomena which may be examined by concentrating on the history of loading. As straining begins, the grains are progressively flattened and the stresses must be increased to continue the process. At a particular value of strain, dynamic recovery and/or recrystallization may begin, and when the rate of softening equals that of hardening, a peak in the true stress–true strain curve is observed which is commonly identified by a pair of variables: the peak strength and strain. For many carbon and alloy steels, dynamic recrystallization then causes further softening. As straining proceeds, the rate of hardening due to straining and dynamic precipitation may equal and/or overtake the loss of strength. The first event results in steady state behavior accompanied by grain refinement, while the latter often causes cyclic hardening/softening, resulting in grain coarsening. A thorough review of the mechanics and metallurgy of hot forming has been presented by Sellars [3]. Hardening and softening during deformation have been discussed by McQueen and Jonas [4] and recently by Jonas and Sakai [5].

Possessing accurate, reliable, and repeatable flow strength data, by itself, is not sufficient if calculations of forces, torques, pressures, and powers during bulk or sheet metal forming processes are contemplated. The material's behavior when subjected to loading must also be represented by an appropriate constitutive equation, giving the strength as a function of other process parameters, such as strain, rate of strain, and temperature. This involves selecting the form of the equation as well as careful nonlinear regression analysis in determining the parameters of that relation. Numerous empirical constitutive equations have been presented in the technical literature. Those specially designed for high-temperature, high strain rate applications include the work of Altan and Boulger [6]; Shida [7]; Gittins et al. [8]; Wusatowski [9], who presents the equation of Ekelund; Hajduk [10]; and Cornfield and Johnson [11]. It may be presumed with some confidence that all these equations are derived from careful experiments conducted on reasonably well-maintained and calibrated testing machines. As well, there appears to be no reason why the regression analyses used in arriving at the reported material parameters should be distrusted.

Comparison of the predictions of Refs 6–11, however, reveals that the information they present is far from convincing. For example, considering AISI 1015 steel at 1200°C subjected to a strain of 50% at a strain rate of 20 s^{-1}, Shida [7] predicts the flow strength to be 97 MPa, Altan and Boulger [6] predict 78 MPa, Cornfield and Johnson [11] give 7 MPa, while Wusatowski [9] prescribes 34 MPa. For other process parameters, the data calculated are similarly inconsistent. It is impossible to ascribe the blame at this point. However, common sense suggests that experimental errors may be at fault.

Consideration of published data on the mechanical response of HSLA steels also reveals contradictions. These may be observed in Table 1, where a compilation of available results on the strength of some niobium-bearing steels is presented. Peak strain, corresponding peak strength, strain rate, temperature, and chemical compositions are given along with the mode of testing and prior solution heat treatment. As is evident, testing techniques, heat treatment, process parameters, and results vary broadly.

Stewart [12] and Armitage et al. [13] use a camplastometer to produce constant strain rate compression; Sankar et al. [14] use a microprocessor-controlled hot torsion testing device, as do Gittins et al. [15]. Tension tests are conducted by Wilcox and Honeycombe [16] and by Maki et al. [19]. Compression of axially symmetrical specimens is carried out by Tiitto et al. [18] and Bacroix et al. [20]. The results of D'Orazio, Mitchell, and Lenard [21] are also given in Table 1.

Parameters used by these researchers differ widely, making direct comparison of the results difficult. Inconsistency of the values is still noticeable. For example, for 0.038% Nb-bearing steel, Tiitto et al. [18] and Maki et al. [19] give data that differ by approximately

TABLE 1—*Compilation of published constitutive results for niobium steel.*

Nb, %	C, %	Mn, %	T, °C	ε, s⁻¹	σ_p, MPa	ε_p	Test	Solution Treatment, °C/min	Reference
0.06	0.16	1.26	900	0.5	154	N/A	Compression	1200°C/60	12
			997	0.5	88	”	”	”	
0.06	0.10	1.65	949	15.0	154	0.25	”	1290°C	13
0.05	0.12	0.94	900	0.1	144	0.75	Torsion	1250°C/30	16
			950	0.1	121	0.65	”	”	
			1000	0.1	104	0.56	”	”	
			900	1.0	185	0.80	”	”	
			1000	1.0	125	0.68	”	”	
0.046	0.05	0.35	1000	7.0	160		Torsion	1100–1300°C/10	15
0.040	0.25	1.15	850	~0.14	79	N/A	Tension	1300°C /30	16
			900	”	62	”	”	”	
			950	”	42	”	”	”	
			1000	”	35	”	”	”	
0.035	0.05	1.25	900	0.00037	73	0.44	Compression	N/A	17
			1075	0.037	63	0.26	”	”	
0.038	0.08	1.25	900	0.006	117	1.20	Compression	1200°C/30	18
			1000	0.006	75	0.30	”	”	
0.038	0.12	1.40	900	0.005	70	0.33	Tension	1100°C/1	19
0.035	0.05	1.25	925	0.014	100	N/A	Compression	N/A	20
				0.0014	75	N/A	”	”	
0.02	0.09	0.90	750	0.2	152	0.54	Compression	1150°C/2	21
				1.0	190	0.62	”	”	
				10.0	206	0.64	”	”	
			850	0.2	110	0.52	”	”	
				1.0	125	0.60	”	”	
				10.0	150	0.63	”	”	
			950	0.2	65	0.50	”	”	
				1.0	80	0.56	”	”	
				10.0	100	0.62	”	”	
0.01	0.16	0.56	1000	7	160	N/A	Torsion	1100–1300°C/10	15

100%. As well, peak strain values appear to depend on the testing method in a significant manner. The only area of agreement concerns the retardation of recrystallization caused by the addition of niobium.

An attempt to determine interlab variability of test results in hot, constant strain rate compression of low-carbon and microalloyed steels was organized recently by the American Iron and Steel Institute. Identical specimens were prepared at one location where the prior thermal treatment was also carried out. Researchers were sent random samples and were requested to follow a carefully prepared procedure. The resulting true stress–true strain curves showed a ±10% variation from laboratory to laboratory as well as from experiment to experiment.

In a thoughtful essay, no doubt, partly inspired by results such as those discussed above, Rowe [22] commented on the availability of analytical techniques of ever increasing accuracy and simplicity for use in the prediction of forces, powers, and velocities in metalforming operations. He further wrote that in those calculations constitutive data of accuracy equal to that of the analytical technique should be used. In the opinion of the present writers that data is extremely difficult to find. This paper describes the understanding of the authors concerning those difficulties.

Results and Discussion

There are three components that need to be considered when developing constitutive data. These are the material to be tested, the testing system and method, and the operator. At this point the infallibility of the operator should be assumed. The remaining two aspects may then be discussed further.

Material

It is the chemical composition of the material that determines its high-temperature strength. In presenting data on flow strength, however, it is not enough to identify the material by its standard designation. Considering low-carbon steels, for example, AISI 1015 may contain 0.13 to 0.18% carbon and 0.3 to 0.6% Mn [23], and, as indicated by Wray [24,25], this causes significant variations of the mechanical, physical, and thermal properties. The exact chemical composition of the sample should be clearly stated in each instance.

The Testing System and Technique

State of Stress—Tension, torsion, and compression testing have and are being used to establish constitutive data [1,2]. Each has its advantages and disadvantages, and each has been shown to produce usable results. The effect of the state of stress during testing on the resulting flow curves, however, has not yet been fully established. On the one hand, Semiatin and Jonas [26] write that for several materials the effective stress-strain data, resulting from compression or torsion testing, agree approximately. On the other hand, they also quote results for Type 304L stainless steel, for which the two types of data differ significantly at room temperatures and at 800°C but agree very well at 1000°C. The authors attribute the differences to the differing rates of dislocation accumulation and texture development present in the two sets of tests. The apparent agreement of the torsion and compression data at the higher temperature needs to be evaluated further, however, in light of the recent work of Jonas and Sakai [5], where peak strains and stresses observed in tension, torsion, and compression indicated significant differences.

Friction at the compression ram-specimen interface also affects the state of stress in the sample. If strains exceed $0.6 \sim 0.8$, the resulting triaxiality of the stresses may cause significant barrelling and the metal's behavior would no longer represent uniaxial loading. These effects may be minimized by efficient lubrication. Use of glass powder-alcohol lubricants and shallow grooves or indentations on the specimen ends to retain them during pressing allows the tests to proceed beyond true strains of 1.5. The stress-strain curves may also be corrected analytically by removing the frictional effects.

Again, reporting the method used to minimize frictional effects is essential.

Stiffness of the Testing System—In any case, whichever testing method is chosen, the effect of the testing machine on the test results also needs to be considered. Of most significance is the deformation of the machine itself, especially when compression testing is considered. Camplastometers usually are of heavy, massive construction, and their distortion during testing is often ignored. Suzuki et al. [27] give the deformation of their camplastometer as 0.084 mm for a force of 147 150 N and calculate the resulting error in strain readings as 0.7% for an original specimen height of 12 mm. This is indeed a negligibly small strain. If the deformed specimen height is taken into account, however, the relative magnitude of the error becomes significant. The flattened sample, subjected to a total true strain of, say, 1.6 will measure 2.42 mm in height. A deflection of 0.084 mm now causes a strain of approximately 3.5%. At a deformed height of 8 mm, corresponding to a true strain of approximately 40%, the discrepancy may amount to a strain of 0.01, which could then lead to a 2.5% error in the total strain. It may be concluded, then, that the stiffness of a camplastometer should be accounted for in the calculations of strains.

Servohydraulic testing machines are, of course, much more flexible, and their distortions are routinely removed from the strain computations. It is hoped that the dynamic effects are also accounted for when test results are reported. The present authors suspect that this is not always the case, and, to underscore the importance of dynamic stiffness calibration, the results shown in Fig. 1 should be considered. There the force-deformation response of a 1332 Model Instron Servohydraulic machine is given. The forces are plotted on the ordinate, and the relative displacements, defined as the difference between the measured displacement and its initial value, are given on the abscissa.

The experiments were conducted at 950°C by compressing the stainless steel loading rams, whose total length is approximately 600 mm. The rate of compression, specifically the load-rate, varied from 0.1 to 10 Hz and is indicated in the figure.

It is evident that dynamic stiffness depends on the rate of loading, and, even if the variations are not excessive, differences of up to 10% are observed.

Considering a force of 60 000 N, this corresponds to change of length of 0.41 mm when

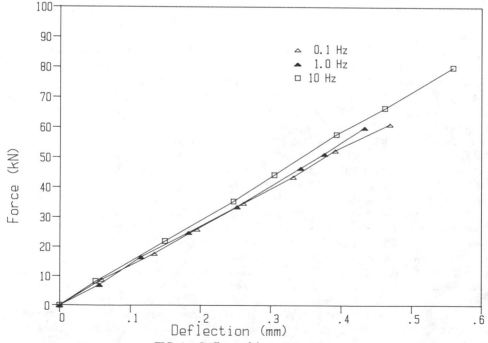

FIG. 1—*Stiffness of the testing system.*

loading at 10 Hz and 0.46 when 0.1 Hz is applied. This is not large, but when errors are cumulative it should be considered.

Testing Technique—Three concepts need to be discussed here. One concerns the pretest thermal treatment, another the ram-sample interface, while the third concentrates on the measurements of temperature.

1. *Pretest Thermal Treatment:* Reference may be made here to the data presented in Table 1. In the second to last column, the pretest thermal treatment, which is necessary to dissolve nitrides and carbides, used by the researchers is given. It is evident that the techniques vary broadly.

To test the contribution of prior heat treatment and/or anisotropy to the strength of a metal, three experiments were conducted (see Fig. 2 for the resulting flow curves). All three tests were conducted at a strain rate of 0.01 s^{-1} and at a temperature of 950°C. A Nb-bearing microalloyed steel containing 0.02% Nb, 0.09% C, and 0.90% Mn was used.

The specimens measured 12.5 mm in diameter and 18 mm in height. Shallow, concentric grooves were machined in the ends of the sample in order to retain the lubricant (Delta Glaze No. 19). No analytical corrections were used.

It is observed that prior heat treatment has a significant effect on the uniaxial flow curves of the material. Solution treatment No. 1 consisted of holding the samples at 1150°C for 30 min and furnace cooling at a rate of 0.3°C/s to the test temperature. The sample was then held for 10 more minutes to reach equilibrium and the compression was performed. Treatment No. 2 included quenching after the 30-min hold period at 1150°C and reheating to 1050°C for 20 min before testing. After furnace cooling to test temperature and holding

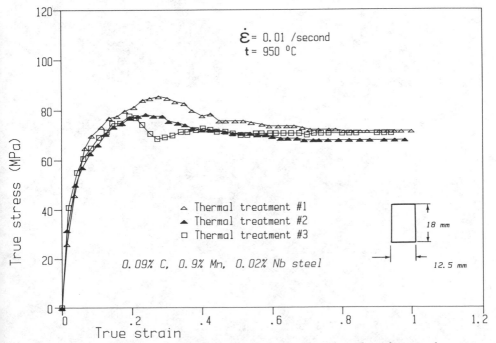

FIG. 2—*Effect of pretest thermal treatment on the flow curves of a niobium steel.*

for another 10 min, the testing was completed. Annealing for 2 h at 1000°C replaced the prior solution treatment for No. 3. The sample was then heated to 1050°C for 20 min, furnace cooled, held there for 10 min, and tested.

All the important parameters observable from the flow curves are affected by the three distinct methods of heat treatment. Peak strengths vary by as much as 10%; peak strains, indicating the beginning of dynamic recrystallization, vary from 0.21 to 0.30. Evidence of cyclic recrystallization is noted to result from treatment No. 3 in addition to some texture-induced strain hardening beyond a true strain of 0.6. Treatment No. 1 and No. 2 have reached steady state conditions at a strain level of approximately 0.75.

The microstructures, all with a magnification of 100, corresponding to the flow curves of Fig. 2, also show some significant variations, as observed in Fig. 3. The micrographs of Fig. 3a and thermal treatment No. 1, Fig. 3b and treatment No. 2, and Fig. 3c and treatment No. 3 belong together. The etchant and etching time were identical for all three cases. It is evident that treatment No. 2 produced the smallest grains, while the two others caused roughly similar development. Treatment No. 1 resulted in a more recrystallized structure than No. 3.

2. *Ram-Sample Interface:* A technique often followed when performing compression tests at high temperatures requires preheating the split furnace to a temperature somewhat above that of the test, placing the specimen with its embedded thermocouple on the ram,

Figure 3a: Treatment #1

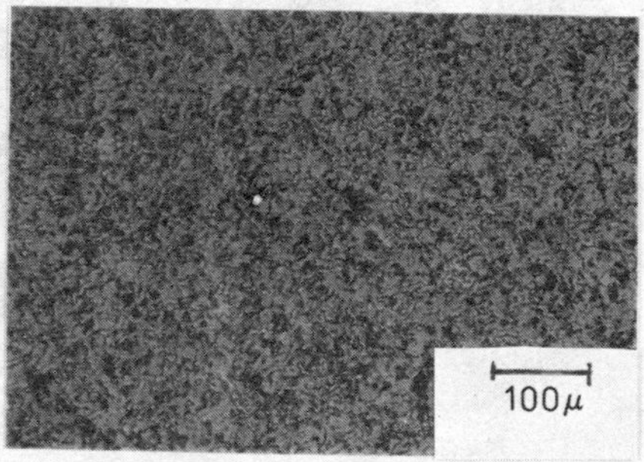

Figure 3b: Treatment #2

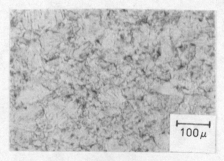

Figure 3c: Treatment #3

FIG. 3—*Effect of pretest thermal treatment on the austenite grains of a niobium steel.*

and monitoring its heating. After the furnace temperature is reached, a few minutes wait is often the rule. The furnace is then opened, the sample is allowed to cool, and when the test temperature is reached the compression is done. The problem here concerns the different cooling rates of the sample and of the ram, including the cooling rates of rams made from stainless steel or Inconel, some of which may also have a ceramic platen. The rams, of course, will cool slower than the samples; further, the steel or Inconel rams will cool faster than those having the ceramic end. If the test is conducted when the sample reaches the necessary temperature and the ram's status is not monitored, the temperature distribution within the specimen may not be uniform and errors may result.

Lubrication at the interface and the preparation of the sample's ends also have an effect on the flow curves, especially at high strains. Corrections for frictional losses are required, and the manner in which this is done varies from researcher to researcher.

3. *Temperature Measurements:* The authors believe that most of the errors, contradictions, and inconsistencies in flow strength data are caused by the temperature measuring system. In order to emphasize the dangers of insufficient care, it is instructive to examine some numerical values of the effect of temperature on strength. For low-carbon steel, an estimate based on the results of Alder and Phillips [28] at low rates of strain indicates a loss of strength of 0.25 MPa/°C. For a 0.09% Nb-bearing microalloyed steel tested in torsion at a strain rate of 7 s^{-1}, the rate of change of mean yield strength with temperature is found to be 0.91 MPa/°C [29]. The results of Suzuki et al. [27] and Altan and Boulger [6] reinforce the above strength/temperature rates. The implication is clear. In the hot forming range, which for steels is in the order of 800 to 1100°C, a $\pm 1\%$ error in temperature may indicate a 10% change in strength.

The most common method of measuring temperatures during testing for hot strength is by thermocouples, which are reasonably inexpensive and easy to use.

Essentially because of their simplicity, it is easy to overlook some important considerations when using one in high-temperature testing. These include the effect of embedding, improper thermal contact, distortion of heat flux in the vicinity of the point of measurement, and response time. In what follows, these points will be discussed in a systematic manner.

(a) *Effect of the embedded thermocouples:* Embedding thermocouples within the plastically worked specimen interrupts the geometry, stress, and strain distribution as well as the temperature distribution. In order to test whether these changes affect the metal's flow strength, uniaxial compression tests were conducted with three specimens of identical geometry (12.5 mm diameter, 18 mm long). In one, two thermocouples with Inconel sheathing of 1.6 mm outside diameter were embedded, one near the top compression platen, the other at the center; in the next two holes were drilled but no thermocouples were placed in them, while the third was left solid. The experiment was carried out at 700°C at a true constant strain rate of 1 s^{-1}. The three flow curves are shown in Fig. 4. As expected, the strength of the solid sample is the largest, followed by the sample with the thermocouples. The weakest is, of course, the specimen with the unfilled holes, having been weakened noticeably by the removed material. The difference in the strength levels exhibited by the solid specimen and the one with the two thermocouples is about 11 MPa, which amounts to about a 5% variation at the 210-MPa stress level. While a 5% difference by itself is not overly significant, the implication of it is quite clear—the stress distribution and hence the strain, rate of strain, and temperature fields have all been affected by the embedded thermocouples. The resulting loss of strength may well contribute to the accumulation of changes and should be taken into account when uncertainties are being investigated.

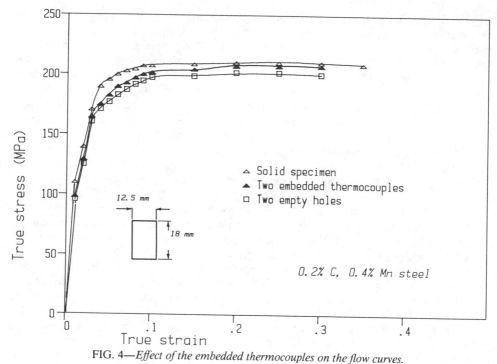

FIG. 4—*Effect of the embedded thermocouples on the flow curves.*

Friction at the ram-sample interface was controlled as in the earlier tests, mentioned above.

(*b*) *Thermal contact:* The errors introduced due to improper or poor thermal contact and heat flow from the thermocouple wire to the object can be serious. An isothermal region should exist along the thermocouple leading up to the object whose temperature is being measured. This is usually accomplished through the use of insulated thermocouples, small thermocouple diameter wire, low thermal conductivity wires, and by supplying additional thermal shielding such as Inconel or stainless steel outer tubing.

Careful installation of the thermocouple in the hole drilled for it would require that the bead make contact with the parent metal. That this has occurred may be easily confirmed by resistance measurements. During deformation, however, especially when finite distortions occur, the specimen, which is of lower strength than the thermocouple sheath, will elongate more than the measuring instrument and contact may be lost. The response as well as the accuracy may be significantly affected by the introduction of air in between the top and the bottom of the hole. Again, as this event may be unavoidable, careful calibration needs to be considered.

(*c*) *Distortion of the heat flux:* Embedding a thermocouple within the body should be such that its tip should attain but not affect the temperature. Since the presence of the hole and the thermocouple in it will interfere with the temperature distribution at the point of interest, only the perturbed temperature will be recorded. If a temperature field is to be monitored, however, thermocouples must be embedded inside and interruption of the continuum is unavoidable. Calibration for the resulting distortion should minimize the errors.

(*d*) *Response time:* During hot forming, significant temperature gradients may develop in a few fractions of a second. The response time of the thermocouple system is critical during these transient type of tests. The rate of response of a thermocouple depends on:

1. The mass of the thermocouple and other physical properties of the sensor.
2. The contact conditions between the thermocouple hot junction bead and the point at which its temperature is being measured.
3. The dynamic temperature dependent properties of the environment.

The response time of the thermocouple is strongly dependent on the heat transfer coefficient of the environment it is placed in. A smaller diameter thermocouple generally provides a faster response than that of similar composition but larger size; it is, however, harder to handle. A compromise is usually required, and the choice is limited—one must accept the fact that the response times of thermocouples are finite.

Conclusions and Recommendations

The difficulties associated with the determination of the metal's resistance to deformation at high temperatures have been discussed. These were identified as those involving the testing system and the method and are given in the list that follows.

1. State of stress during testing.
2. Stiffness of the testing system.
3. Testing technique, including: (a) pretest thermal treatment; (b) ram-sample interface, which involves the heat transfer characteristics of the contacting materials, lubrication, and sample geometry; (c) temperature measurements, involving the effect of embedding the thermocouples in the samples, thermal contact, distortion of the heat flux, and response time.

In considering these phenomena it is apparent that if no standard procedures are followed, no comparable flow curves will be obtained. The following is then recommended.

For the determination of constitutive data, the pretest thermal treatment and austenite grain size, the sample size and end geometry, the material of the loading ram, the precise time schedule of the experiment, the lubricant, the type of thermocouple, the measuring system, the type of test, and the material's chemical composition should always be reported.

Finally, it was not possible to determine the magnitude of an acceptable error band for the flow curves. In considering the above, however, a $\pm 5\%$ inter- and intralaboratory variability of test results does not appear excessive.

Acknowledgments

The authors are grateful for the financial assistance received from the Natural Sciences and Engineering Research Council of Canada and NATO under Special Research Grant No. 390/83.

References

[1] Alexander, J. M., *W. Olzak Memorial Volume,* H. Sawczuk, Ed., Elsevier Applied Science Publishers, Ltd., in press.
[2] Lenard, J. G., *Journal of Engineering Materials and Technology,* Vol. 107, 1985, p. 126.

[3] Sellars, C. M., *Proceedings,* International Conference on Hot Working and Forming Processes, The Metals Society, Sheffield, England, 1979, p. 3.

[4] McQueen, H. J. and Jonas, J. J. in *Metalforming—Interrelation Between Theory and Practice,* A. L. Hoffmanner, Ed., Plenum Press, New York, 1971, p. 393.

[5] Jonas, J. J. and Sakai, T. in *Deformation, Processing and Structure,* G. Krauss, Ed., American Society for Metals, Metals Park, OH, 1984, p. 185.

[6] Altan, T. and Boulger, F. W., *Journal of Engineering for Industry,* Vol. 95, 1973, p. 1009.

[7] Shida, S., Hitachi Research Laboratory Report, Hitachi Ltd., Tokyo, 1974, p. 1.

[8] Gittins, A., Moller, R. H., and Everett, J. R. *BHP Technical Bulletin,* Vol. 18, 1974, p. 1.

[9] Wusatowski, Z., *Fundamentals of Rolling,* Pergamon Press, Oxford, 1969.

[10] Hajduk, M., et al., BISITS 11281, British Iron and Steel Industry Translation Service, London, 1973, p. 1.

[11] Cornfield, G. C. and Johnson, R. H., *The Journal of the Iron and Steel Institute,* Vol. 211, 1973, p. 567.

[12] Stewart, M. J., CANMET Report ERP/PMRL-75-18(J), Energy, Mines and Resources, Canada, 1975.

[13] Armitage, B., McCutcheon, D. B., and Newton, L. D., *Proceedings,* 18th Conference on Mechanical Working and Steel Processing, Harvey, IL, 1976, American Society of Metals, Metals Park, OH, p. 13.

[14] Sankar, J., Hawkins, D., and McQueen, H. J., *Metals Technology,* Vol. 6, 1979, p. 325.

[15] Gittins, A., Everett, J. R., and Tegart, W. J. M., *Metals Technology,* Vol. 4, 1977, p. 377.

[16] Wilcox, J. R. and Honeycombe, R. W. K., *Metals Technology,* Vol. 11, 1984, p. 217.

[17] Sakai, T., Akben, M. G., and Jonas, J. J., *Proceedings,* International Conference of Thermomechanical Processing of Microalloyed Austenite, Pittsburgh, AIME, New York, 1981, p. 237.

[18] Tiitto, K., Fitzsimmons, G., and DeArdo, A. J., *Acta Metallurgica,* Vol. 31, 1983, p. 1159.

[19] Maki, T., Akasaka, K., and Tamura, I., International Conference of TPMA, Pittsburgh, 1981, AIME, New York, p. 217.

[20] Bacroix, B., Akben, M. G., and Jonas, J. J., *Proceedings,* International Conference of TPMA, Pittsburgh, 1981, AIME, New York, p. 293.

[21] D'Orazio, L. R., Mitchell, A. B., and Lenard, J. G., *Proceedings,* HSLA 85, Beijing, People's Republic of China, 1985, p. 189.

[22] Rowe, G. W., *Journal of Mechanical Working Technology,* Vol. 11, 1985, p. 1.

[23] *Engineering Properties of Steel,* Harvey, P. D., Ed., American Society for Metals, Metals Park, OH, 1982.

[24] Wray, P. J., *Metallurgical Transactions A,* Vol. 13A, 1982, p. 125.

[25] Wray, P. J., *Proceedings,* Modeling of Casting and Welding Processes, Rindge, NH, 1980, AIME, New York, p. 245.

[26] Semiatin, S. L. and Jonas, J. J., *Formability and Workability of Metals,* American Society for Metals, Metals Park, OH, 1984.

[27] Suzuki, H., Hashizume, S., Yabuki, Y., Ichihara, Y., Nakajima, S., and Kenmochi, K., University of Tokyo Report, Vol. 18, No. 3, 1968.

[28] Alder, J. F. and Phillips, V. A., *Journal, Institute of Metals* (London), Vol. 83, 1954–55, p. 80.

[29] Everett, J. R., Gittins, A., Glover, G., and Toyama, M., *Proceedings,* International Conference on Hot Working and Forming Processes, C. M. Sellars and G. J. Davies, Eds., The Metals Society, Sheffield, England, 1979, p. 16.

Kiyoshi Taniuchi[1]

Simple Stress Sensor: Utilizing of Stretcher Strains

REFERENCE: Taniuchi, K., **"Simple Stress Sensor: Utilizing of Stretcher Strains,"** *Factors That Affect the Precision of Mechanical Tests, ASTM STP 1025,* R. Papirno and H. C. Weiss, Eds., American Society for Testing and Materials, Philadelphia, 1989, pp. 217–232.

ABSTRACT: This article concerns an attempt to utilize stretcher strains to determine the stress concentration factor in the elastic-plastic state of steel strip specimens having semicircular notches. The dangerous section of steel gear tooth profiles is also brought under examination. In both cases, the utilization of stretcher strains provides successful results. For this undertaking, the observation of the striped pattern was made solely by the naked eye, without the use of a measuring instrument.

The examples shown will illustrate that the use of stretcher strains is a simple and convenient stress sensor.

KEY WORDS: stress, sensor, stretcher strains, elastic-plastic, stress concentration, dangerous section, gears, experimental stress analysis

For the sake of pressing thin steel sheets with no stretcher strains, material science has done much research on the nature of stretcher strains [1]; however, there has not been any substantial research on the feasibility of using stretcher strains as a technique for the study of notched material strength [2,3], and as yet there has not been a single report which has relied solely on the measurement of stretcher strains.

It is generally known that stretcher strains correspond to yield stress in that yield stress can be discerned by the naked eye as a striped pattern [1]. This suggests that stretcher strains enable one to visualize the elastic-plastic state of steel and thus can be used as a kind of stress sensor.

An examination was undertaken of the form of striped patterns that appear on the surfaces of specimens of carbon steel strips. The carbon steel strips each had a pair of facing semicircular notches, and with an increase in tensile load the striped patterns appeared at the notches. A similar examination was made of the striped patterns which developed when rectangular bars were bent. The results of these investigations made it possible to determine the stress concentration factor for the elastic-plastic state of semicircular notched pairs and for the dangerous section of tooth profiles. The results are presented in this report.

Strip Specimens and Experimental Method

Strip Specimens

In Fig. 1, the form and dimensions of a specimen are shown. It is a JIS No. 13B specimen strip in which a pair of semicircular notches were cut. Five variables ranging from 1.25 to

[1] Lecturer, Meiji University, Faculty of Engineering, Kawasaki-shi, Kanagawa-ken, Japan 214.

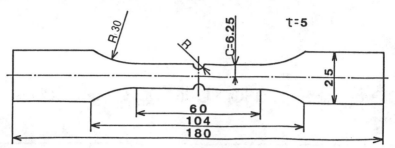

FIG. 1—*The dimensions of a notched strip tension specimen.*

3.25 mm were empirically selected for the radii R of the notches. Figure 2 is an enlarged photograph of the notched parts. The bottoms of the notches were finished into smooth semicircular curves. The specimens were annealed after the machining was completed. The finishing was subsequently refined with emery papers and finally lapped with chromium oxide. At this point its Vickers hardness number was measured and determined to be 125. Table 1 shows the chemical composition of the specimens. The optical microscope photograph of the structure given in Fig. 3 made it possible to determine that the crystal grain size was 16 μm.

Test Gear

The experiment was carried out with a standard 4-mm module spur gear with 30 teeth, a pressure angle of 20°, and a gear width of 10 mm. The accuracy inspection of the gear rated as JIS Class 4. Table 2 shows the chemical composition of the material. Figure 4 is the optical microscope photograph of the structures, and the crystal grain size was 30 μm.

The test gear was annealed after hobbing, and the flank of the teeth was further lapped with chromium oxide abrasive.

Experimental Method

Test of Notched Strip Specimens—An Instron-type material testing machine was used to test the tensile load and determine the load-elongation curves. While the tensile load was

(Scale mark indicates 1.5mm)

FIG. 2—*Detail of the notched area of a strip tension specimen.*

TABLE 1—*Chemical composition of the notched tension specimens, %.*

C	Si	Mn	P	S
0.12	0.21	0.55	0.019	0.013

being applied, the surfaces of the specimens were carefully observed by the naked eye. The crosshead speed of the testing machine was set at 0.5 mm per min. This value was decided upon to make the phenomena arising on the specimen surface comfortably observable by the naked eye.

Careful examination was made of how the striped patterns appeared and propagated on the surfaces of the notched parts as the tensile load was increased. When it was determined at what degree of tensile load the striped patterns began to appear, the curves of a self-recorder were used in conjunction with the data given by observing the specimen surfaces with the naked eye.

Measurement was made of the surface roughness at the places where the striped patterns on specimen surfaces appeared, and the measured profile indicated the maximum height R max.

Bending Test of Gear Teeth—Figure 5 is a sketch of the experimental setup. The test gear (1) was engaged with another gear (2); a torque was then applied to the shaft of the first gear (1), but the wedge action of a roller (5) was used to prevent the movement of the second gear (2). Thus, a force was generated to act jointly on both engaged teeth. In order to limit the experiment to the engagement of only a pair of teeth, as is considered fundamental in the strength analysis of gear teeth, of the 30 teeth of the second gear (2), both teeth adjacent to the engaged tooth were cut off. The tooth width of the second gear (2) was 40 mm, thus four times the 10 mm width of the tooth of the first gear (1). The load W was applied by steadily pulling by hand the end of the lever (3) through a spring balance as shown in the figure.

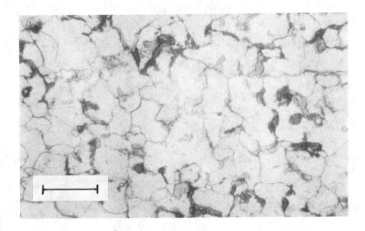

(Scale mark indicates 30μm)

FIG. 3—*The microstructures of the notched strip tension specimens.*

TABLE 2—*Chemical composition (test gears), %.*

C	Si	P	Mn	S
0.19	0.29	0.022	0.34	0.010

After both gears were set in the prescribed state, the flank at the engaged tooth of the first test gear (1) was brought under observation by the naked eye as the load W was steadily applied. The value of W ranged from 147 to 215 N. At the same time, a moment arm of 1120 mm in length and the 120-mm pitch circle diameter of the gear exerted a force on the tooth of the gear ranging from 2940 to 4312 N.

Results and Examination

Notched Specimens

Striped Patterns Under Various Loads—Observation was made of the change of surface patterns of the specimen relative to the increase of tensile load. Figure 6 indicates the results. When the load had reached a certain value, $P(a)$, a leaf-bud–like striped pattern appeared at the bottom of the notch on one side as shown in Fig. 6(a). A striped pattern of similar form appeared subsequently in the adjoining region. As the load increased, the patterns continued to grow toward the bottom of the notch on the other side and after a while crossed over the area between the bottoms of the notches on both sides and mutated into the state shown in Fig. 6(b). In the next stage, as shown in Fig. 6(c), the patterns covered the whole notched region. The striped pattern thereafter gradually propagated on the parallel surface of the specimen toward its shoulders as shown in Fig. 6(d), and, finally, all the striped patterns which had so far appeared instantly changed to pear-skin patterns unlike those observed so far. These patterns also changed little by little proportionally to

(Scale mark indicates 30μm)

FIG. 4—*The microstructure of test gears.*

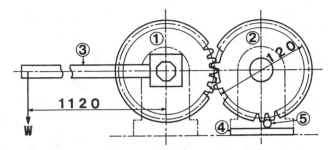

FIG. 5—*The apparatus for testing gears.*

the increase of load until the specimen broke at its notched part. Figure 6(*e*) shows its state at the moment after the specimen was pulled apart.

Figure 7 shows the profile of the measurements of the surface roughness of the striped patterns in Fig. 6(*c*). Figure 8 displays for the sake of comparison the profile of the measurements of a specimen surface before tensile load was applied. The profile of measurements displayed a concave form instead of striped patterns. The *R* max value was 45 μm.

The results in Fig. 6 show that the change of the striped patterns appearing on specimen surfaces in conjunction with the increase of load is gradual and regular. The process of this change can easily be observed by the naked eye.

Load-Elongation Diagrams—Figure 9 is a load-elongation diagram derived from a specimen at *R* = 3.0 mm. The symbols $P(a),P(b)$ in the figure show the load corresponding to the respective striped patterns (*a*),(*b*) shown in Fig. 6. The $P(a)$ that indicates the load at the moment when the striped pattern of Fig. 6(*a*) appears is within the range where the load and the elongation are still proportionally related. At this moment the graph alone is incapable of indicating the position of $P(a)$; unless the data of observing the specimen surface are supplemented, the position of $P(a)$ remains indeterminate. $P(b)$ indicates the load at the moment that the striped pattern of Fig. 6(*b*) crossed over the area between the bottoms of the notches of each side, and in this vicinity, the line in the graph begins to curve slightly.

It is easy to mark the positions on the load-elongation diagram which correspond to the gradual mutation of the striped patterns. Utilizing these data, it becomes feasible to estimate the degree of tensile stress exerted on the specimen simply by observing with the naked eye the nature of the striped patterns which have appeared on the surface of the specimen.

Estimation of Stress at the Bottoms of Notches—Hypothetically, the observation of stretcher strains is an efficacious means to measure tensile load if a clear relation can be established between the appearance of the striped patterns at the bottoms of the notches and the degree of the tensile stress exerted in the same location.

According to Ogura et al., the master curves can be interpreted to reveal the stress concentration factor α_σ of the elastic-plastic state. The master curves are constructed from the nominal stress in the smallest cross section between the bottoms of notches and the stress concentration factor α [4,5]. As a result, the maximum stress, that is, the stress at the bottoms of notches, can be estimated by

$$\sigma \max = \sigma_n \times \alpha_\sigma \qquad (1)$$

where σ_n = the nominal stress at the minimal cross section. The stress at the bottom of the notches was estimated using this method.

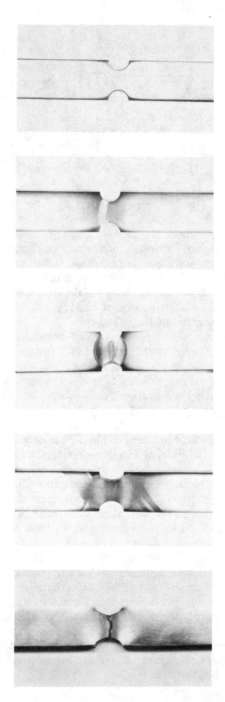

(R:3.0mm)

FIG. 6—*The propagation of stretcher strains in notched strip tension specimens.*

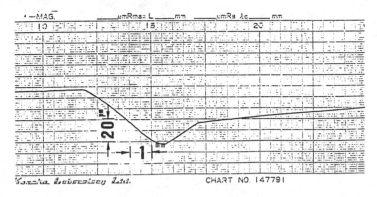

(R : 3.0mm)

FIG. 7—*The profile of the measurements of the surface roughness of stretcher strains on a notched strip tension specimen.*

The $P(a)$ load at the moment when striped patterns appeared at the bottoms of the notches of specimens having a notch radius R of 3.0 mm was 7820 N. At the same time the nominal stress σn was 241 MPa. According to Ref 6, the stress concentration factor α for $R = 3.0$ mm is 1.69. Using the master curves proposed by Ogura et al. [5], the stress concentration factor for these conditions, particularly for the case of low work hardening, was determined to be 1.25. Thus, through Eq 1 the estimated value of the stress at the bottom of the notches was determined to be 301 MPa. At this point, since the experimental values dispersed when the striped patterns began to appear, the estimated value of the stress took on a certain degree of latitude.

When a tension test was performed on a JIS No. 13B specimen of a material having the same chemical composition as the material of the specimens of Table 1, it was determined that its yield stress was 295 MPa. This value may be regarded as the approximate estimated value of the stress at the bottoms of the notches. The results of these calculations support the postulate that stretcher strains appear at the same place where yield stress is exerted at the bottoms of notches.

The place where yield stress acts in a specimen which has been finished smoothly can

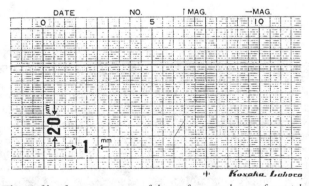

FIG. 8—*The profile of measurements of the surface roughness of a notched strip tension specimen.*

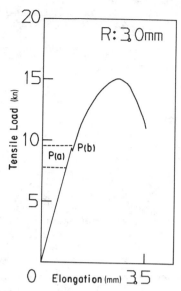

FIG. 9—*The load elongation diagram of a notched strip tension specimen.*

be detected by the presence of striped patterns. A remarkable feature of this method is that from the moment that the striped patterns appear no instrument for measurement is needed to detect yield stress.

Notch Radius and Beginning of Striped Pattern Appearance—The use of stretcher strains as a sensor of yield stress made it possible to determine the relation between the change of notch radius and the load at which striped patterns began to appear.

Figure 10 is an arrangement on the ordinate of the load $P(a)$ at which the striped patterns began to appear at the bottoms of the notches, while the corresponding locations on the abscissa indicate the radio R/b of the notch radius R to the half-width of the parallel parts b of the specimens. The plots ran regularly, lowering rightward, and as R/b became larger, $P(a)$ became smaller. When an empirical formula was determined from this result, it resulted in the following linear equation

$$P(a) = (-1.78 \, R/b + 1.64) \times 10^4 \qquad (2)$$

The correlation coefficient of this equation was -0.955. The straight line in Fig. 10 represents Eq 2. The hatched part furthermore indicates a range of 95% as a confidence limit.

The load $P(a)$ at which striped patterns began to appear at the bottoms of the notches was determined to be the linear function of R/b. The stress at the bottoms of notches can be estimated from the value of $P(a)$. The ascertainment of the stress concentration factor in the elastic-plastic state essentially depends on the experimental values of the stress at the bottoms of the notches.

It was confirmed that stretcher strains provide a reliable method for the experimental analysis of the stress concentration in the elastic-plastic state. In this method, the only preparation needed is simply to smoothly finish the surfaces of the specimens. Accordingly, it is very simple and convenient as compared with the various other methods of experimental stress analysis generally used at present.

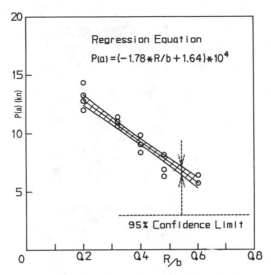

FIG. 10—*The relation between* P(a) *and* R/b.

Stretcher Strains at Gear Tooth Roots

Form of Striped Patterns—Figure 11 [7] is an enlarged illustration of the striped patterns which appeared, grew, and propagated on the flank at the tooth root of the test gear when it was subjected to a torque. As the torque increased, the striped patterns changed in succession from (a) to (b) in Fig. 11. The arrow marks in these figures indicate the direction of the force applied to the tooth.

First, the striped patterns arose on the corner curves on both sides of the tooth root. As seen in Fig. 11(a), the patterns took the form of leaf buds, and their tips pointed towards the center line of the tooth profile. As the torque increased, the respective tips of the patterns on both sides approached each other. At the same time their width increased and the patterns transformed into those shown in Fig. 11(b). The patterns which appeared first continued growing, and subsequently a number of patterns similar in form appeared adjacently to the initial patterns and began to develop in the same sequence. At the same time the flank of the tooth root mutated as shown in Fig. 11(c). The propagation of the striped patterns going toward the bottom side of the tooth stopped soon thereafter, but the striped patterns going toward the side of the tooth tip spread successively through the adjacent regions, and soon the flank of the tooth mutated into the form shown in Fig. 11(d).

The striped patterns on the flank of the tooth stayed the same even after the load applied to the gear was removed. In this way it is possible to gradually correlate what striped patterns correspond to what intensity of force applied to the tooth.

The observation of stretcher strains enables the naked eye to continuously trace the elastic-plastic behavior of a tooth gear.

The record in Fig. 11 correlates well with the results of Ishikawa [8], who examined the yield process of gear teeth by the Schlieren method.

The patterns given in Fig. 11 reveal that the forms of the patterns on the opposite sides of a tooth look different from each other. In order to account for the cause of this phenomenon, the record in Fig. 12 was constructed to indicate the surface roughness at the corners of the tooth root in Fig. 11(c). The arrow marks at the corners of the tooth profiles recorded

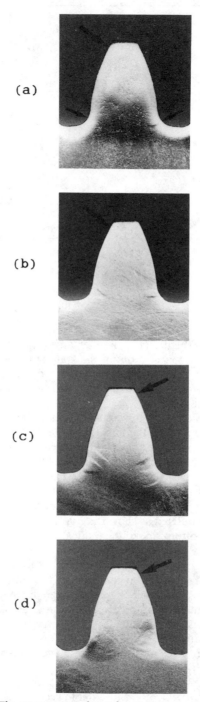

(a)

(b)

(c)

(d)

FIG. 11—*The propagation of stretcher strains on a gear tooth* [7].

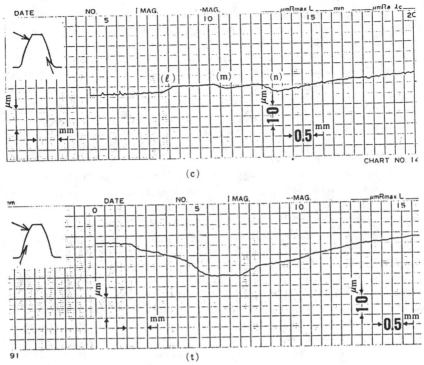

FIG. 12—*The profile of measurements of stretcher strains at the roots of a gear tooth.*

in this diagram show the direction in which the probe of the measuring instrument moved, and the arrow marks pointing toward the tooth tips represent the direction of the force applied. Figure 13 is the profile of the measurements of the surface roughness of the tooth root flank of the gear before being subjected to a load.

When the record in Fig. 12 was observed while using that in Fig. 13 as a reference, it was found that the form on the side of the tension became concave, while the side of the compression protruded. It is conjectured that the discrepancy of the visual observation in the record of Fig. 11 was mainly a result of the difference of the concavity and protrusion of the profile of measurments of the surface roughness.

The places l, m, and n of the record in Fig. 12 correspond to the striped patterns in Fig.

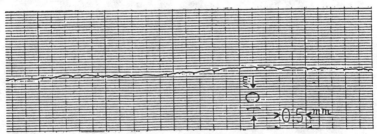

FIG. 13—*The profile of measurements of the roots of a gear tooth.*

11(c). The width of all the protrusions were measured to be within 0.5 mm. It was found that when the position was determined by utilizing stretcher strains, the accuracy of the reading was also within 0.5 mm.

The Position of the Striped Patterns Appearing at the Tooth Root—Figure 14 [7] is formed when the right half of the photograph shown in Fig. 11(a) and the left half of a tooth profile curve are juxtaposed on each side of the center line of the tooth profile. The left half is the figure in which the dangerous section was recorded in a drawn tooth profile curve [9] by the 30° tangent method. Figure 14 shows that the position of the striped patterns and the dangerous section of the tooth profile are mutually symmetrical in relation to the center line of the tooth profile. This confirms that the positions of the striped pattern and of the dangerous section coincide.

The observation of stretcher strains enables the naked eye to discern the dangerous section at the tooth root of an actual gear. Based on the record given in Fig. 12, this observation may be counted upon to have an accuracy latitude of 0.5 mm.

Striped Patterns in Bending Specimens—The stress acting at the tooth root of a gear subjected to a torque is almost a bending stress. When utilizing the observation of stretcher strains for the analysis of gears, it is necessary to integrate the data on the relation of bending stress to the appearance and propagation of the striped patterns.

Rectangular bars having the dimensions of 18 × 10 × 200 mm were made of the materials indicated in Table 2, similar to that of the test gear, and they were subjected to a bending test by two-point loading and two-point support. Figure 15 [10] is a record of the striped patterns appearing on the sides. Many striped leaf-bud patterns appeared side by side at the upper and lower edges and moved toward the middle of the rectangular bar. The form of the profile of measurements of these patterns showed that the side which was

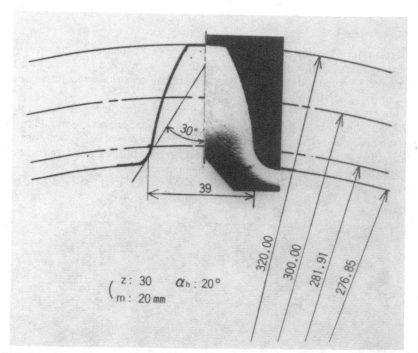

FIG. 14—*The relation between the dangerous sections and the generating positions of the stretcher strains of a gear tooth* [7].

FIG. 15—*The stretcher strains on a rectangular bar at the moment of being bent* [10].

acted upon by tensile stress became concave, while the side which was acted upon by compressive stress formed a protrusion [*10*]. Comparison of the striped patterns of Fig. 15 and Fig. 11(*c*) showed that they were similar to each other. The data most essential for the investigation of the behavior of the striped patterns on the tooth root flanks are the mechanical characteristics of the striped patterns that appear on the surfaces of the bending specimens.

Stretcher strains appear as striped patterns on the smooth surfaces of steel specimens subjected to yield stress. The stretcher strains have been variously named as: Lüders lines, Hartmann lines, Lüders bands, and Probert lines. However, stretcher strains are invariant in regard to whether the stress is tensile or compressive; therefore, perhaps "yield stress pattern" or "yield stress stripe" might be the most appropriate names for this phenomenon [*10*].

Discernment of Dangerous Sections

Figure 16 delineates where striped patterns arose for the first time on a JIS No. 13B specimen. Striped patterns first appeared at the boundaries of the parallel edges and at the shoulder radii. Stress concentration arises in the following places.

The striped patterns of Fig. 6(*a*) began to appear on the smallest cross section at the notched part, and this was also the location of stress concentration.

In the case of a gear tooth profile, as seen in Fig. 14, the striped patterns appeared at the dangerous section.

The place weakest in strength reaches the yield stress first, and at that location the striped pattern appears to reveal shear stress. This is because the striped patterns are the result of yielding along shear planes and occur at the area of maximum shear stress [*11*]. The dan-

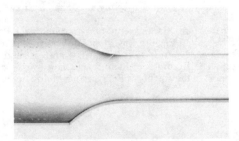

FIG. 16—*The generating position of the stretcher strains on a stepped strip tension specimen.*

gerous section and the position where the striped pattern first appears may be regarded as the same. There may nevertheless be other points of even higher stress.

By observing the striped patterns of stretcher strains with the naked eye after having finished the surfaces smoothly, it is possible to locate the dangerous section of a member [12].

Conclusion

Considering that the characteristics of stretcher strains corresponding to yield stress perform the role of a stress sensor, determination was made of the stress at the bottoms of semicircular notches in steel strips by utilizing the striped patterns in an actual example as a criterion. Striped patterns were also used to discern the dangerous section of a gear tooth profile and the dangerous section of stepped strips. In all cases the results were successful. The observation of the striped patterns also enables one to visualize the behavior of steel in the elastic-plastic state.

The experiments confirm that stretcher strain phenomena can be utilized as a kind of stress sensor. The method of using this sensor is very simple and convenient, and it would be desirable to characterize it as a technique for finding the location of the highest in-shear stresses in certain annealed carbon steel strip specimens. Since there are perhaps already too many names for this phenomenon, as mentioned before, it may be more intelligible and appropriate to replace the name "stretcher strains" and its other epithets with one of two simpler names: "yield stress pattern" or "yield stress stripe."

References

[1] Sudo, E., *Stretcher Strains,* Nihon Kinzoku Gakkai, 1970.
[2] Maekawa, I. and Sakurai, M., *Transactions of the Japan Society of Mechanical Engineers,* Vol. 35, No. 271, March 1969, pp. 482–490.
[3] Maekawa, I. and Matsuyama, Y., *Transactions of the Japan Society of Mechanical Engineers,* Vol. 37, No. 300, August 1971, pp. 1483–1491.
[4] Ogura, K., Oji, K., and Takii, H., *Transactions of the Japan Society of Mechanical Engineers,* Vol. 41, No. 341, January 1975, pp. 87–95.
[5] Ogura, K., Miki, N., and Oji, K., *Transactions of the Japan Society of Mechanical Engineers,* Vol. 47, No. 413, January 1981, pp. 55–62.
[6] *JSME Handbook for Mechanical Engineers, A Fundamental. A4; Strength of Materials,* 1984, p. 97.
[7] Taniuchi, K. and Satoh, T., "Results of the Observation on the Texture of Stretcher Strains at the Base of Gear Teeth," *Memoirs of the Institute of Science & Technology,* Meiji University, Vol. 24, No. 10, 1985.
[8] Ishikawa, N., *Journal of the Japan Society of Mechanical Engineers,* Vol. 52, No. 371, November 1949, pp. 398–402.
[9] Taniuchi, K. and Hayashi, K., *Design & Drafting,* the Japan Society for Design & Drafting, Vol. 8, No. 32, 1972, pp. 8–13.
[10] Taniuchi, K., "Stretcher-Strain Markings on Low-Carbon Steel Bars Caused by Bending," research report of the Faculty of Engineering, Meiji University, No. 50, 1986, pp. 15–21.
[11] Timoshenko, S., *Strength of Materials, Part I,* 3rd ed., Van Nostrand, New York, 1955, p. 39.
[12] Taniuchi, K. and Satoh, T., "Examples of Identifying Dangerous Sections with Stretcher-Strain Markings," research report of the Faculty of Engineering, Meiji University, No. 48, 1985, pp. 35–43.

DISCUSSION

T. G. F. Gray[1] (written discussion)—I found Professor Taniuchi's findings extremely interesting and wonder if he has determined the uniaxial lower yield strength of his test material in a *plain* form, that is, in a specimen with no stress concentration present. This value might then be compared with the values obtained with notches. My reason for asking is that some years ago I was faced with a problem of yielding in a cracked tension specimen where it was clear that the effective yield strength of the material in the notched configuration was significantly and consistently higher than the yield strength in a plain specimen. The material type and specimen finish were such that stretcher strain markings were visible, as in Professor Taniuchi's work.

To resolve this problem of disagreement in yield value in the two configurations, I argued that in the plain case, a yield band may form at one spot and continue at that level of load unhindered; whereas in the notched case, yielding is triggered at various stress levels in a broader zone local to the notch. In the latter case, the immediate interaction between dislocations in this zone may lead to a certain elevation of apparent flow strength. My own solution to the problem was to attempt to simulate the dislocation tangling process which I suspected to be occurring in the notch zone by applying a ratchetting strain-controlled program to specimens. The effective yield was then deemed to be the upper envelope of the cyclic load/extension curve. This method produced a significantly higher apparent yield strength (20% approximately) with much less variation from test to test.

One conclusion of this kind of finding might be that the kinds of test we normally do on plain specimens may not be too relevant to real applications where there are stress concentrations. Maybe new standards are required!

K. Taniuchi (author's closure)—The yield strength indicated by the experiments I performed have valid application only to a JIS 13B specimen strip. The specimens were annealed after the machining was completed.

I would also like to mention that Fig. 16 in fact shows the original condition of the specimen prior to when the stretcher strains appear at the location of stress concentration.

I feel that it would be extremely interesting to undertake the investigation that you suggested on the variation of specimen strength according to form, and I plan to prepare an experimental procedure for exploring this sort of variation.

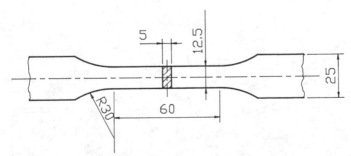

FIG. B—*The dimensions of a JIS No. 13B specimen.*

[1] University of Strathclyde, Dept. of Mechanical & Process Engineering, Division of Mechanics of Materials, James Weir Building, 75 Montrose Street, Glasgow GI 1XJ, Scotland.

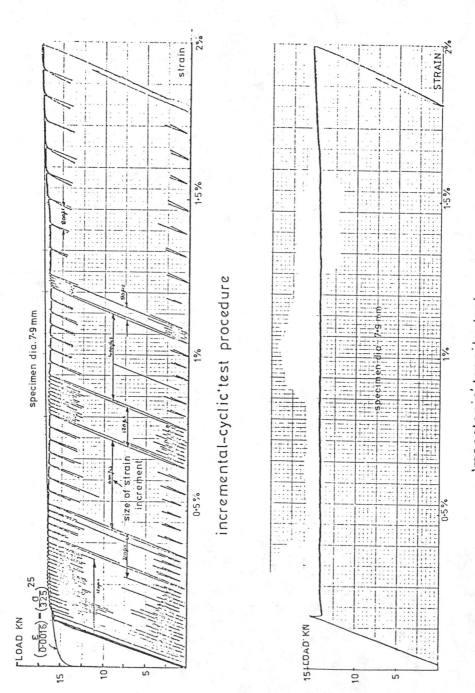

FIG. A—*Incremental-cyclic test procedure and monotonic tension test.*

Jerry L. Lower[1] and Howard C. Price[1]

Weight Loss Technique for Measurement of Wear of Polymeric Orthopedic Implants

REFERENCE: Lower, Jerry L., **"Weight Loss Technique for Measurement of Wear of Polymeric Orthopedic Implants,"** *Factors That Affect the Precision of Mechanical Tests, ASTM STP 1025,* R. Papirno and H. C. Weiss, Eds., American Society for Testing and Materials, Philadelphia, 1989, pp. 233–239.

ABSTRACT: Low wear properties of ultra-high molecular-weight polyethylene (UHMWPE) and carbon reinforced polyethylene prosthesis require 0.1-mg accuracy for a valid quantitative wear test. Buoyancy of air displaced by the volume of the wear specimen will have an influence on the specimen's weight. Density variation of air due to barometric pressure, temperature, and relative humidity during a wear test will alter the succeeding weight measurements. The average amount of wear debris is very small compared to the total mass of the wear specimen, approximately 0.005 to 0.05% or 1 to 2 mg per million cycles for a 4000 to 22 000-mg component, respectively. Displaced air mass variation for a volume equivalent to a wear component over a test period can be as high as 1 mg. Buoyancy compensation should be considered for addition to ASTM F 732 (Practice for Reciprocating Pin on Flat Evaluation of Friction and Wear Properties of Polymeric Materials for Use in Total Joint Prostheses) and any test procedure utilizing weight loss measurement for wear.

KEY WORDS: buoyancy, wear, weight, density, barometric pressure, temperature, relative humidity, wear debris, mass, volume

Quantitative wear measurement of polymeric total joint prosthesis can be improved. Different methods have been devised for quantitative wear studies, but each method has advantages and disadvantages. One such method is the weight loss technique as used in ASTM F 732 [1]. The loss of weight over the time of the test after correction for fluid sorption is considered to be the wear.

There are three specimen-related factors which influence the total weight loss due to wear of a polymer component. The first factor is fluid weight gain. McKellop et al. [2] found that fluid weight gain is a significant percentage of the total weight change (absorption uptake by the polymer specimen). Weight gain is considered constant after 30 days, and the slope of the increase is calculated to Day 45 [3]. For long wear tests (50 to 100 days), it is best to use control samples soaked 100 days or more. Control samples are then loaded the same as the test sample without motion. This will determine fluid absorption under load. The second factor which can influence weight change is inconsistent cleaning and drying procedures. This is another major source of error in the incremental weight loss due to wear. A cleaning and drying protocol, as in the Appendix, must be strictly adhered to. The implant must be carefully removed from the test fixture and cleaned, ensuring no removal of the substrate material, that is, removing only the debris and the lubricating fluid. The last major factor which can significantly influence the observed weight of a component part is the ambient buoyancy variation.

[1] Senior research engineer and research group manager, respectively, Zimmer, a division of Bristol Myers, Warsaw, IN 46580.

The buoyancy force of air is described by Archimedes' principle "the net vertical force of a fluid on a foreign body is equal to the magnitude of the weight of the displaced fluid." Archimedes' principle holds true for compressible fluid as well as for noncompressible liquids [4]. Displaced air mass variation for a volume equivalent to a knee polymer component over a six-month period can approach 1 mg (Table 1). The density variation of air from the beginning to the end of the test will alter the observed weight of the specimen.

Procedure for Buoyancy Variation

Seventeen sterilized compression molded UHMWPE pin cylinders (35 mm long by 9 mm in diameter) for use on an ASTM F 732 type reciprocating pin on flat test were soaked in filtered-sterilized calf serum at 37°C for 140 days. The serum (supplied by Hazleton Research Products, Lenexa, KS) had 0.3% sodium azide added to control bacteria growth. The pins were removed, cleaned, and weighed periodically ten times using the cleaning procedure in the Appendix. Standard F 732 Section 6.2.5 indicates that the analytical balance should have an accuracy of 10 μg. The barometric pressure, temperature, and relative humidity were recorded at the same time the pins were weighed.

Results

The average of 17 pins was determined and the mean and standard deviation calculated (Table 2). All pins exhibited a rapid weight gain for the first 30 days and a gradual long-term linear trend similar to the study by I. C. Clarke and W. Starkebaum [3]. The average weight gain versus the time the sample was soaked is shown in Fig. 1. The weight change due to the buoyancy effect can be calculated by determining the weight of the air displaced

TABLE 1—*Weight change of a volume of air in tribology laboratory from 3/14/86 to 9/2/86.*

	Time	Temperature, °F/°C	Humidity, %	Barometric Pressure, in. Hg/mm Hg	Air Displaced[a], 21.8 cc-16 mm Knee	
					Moist Air, Mg	Dry Air, Mg
Max barometric pressure Min humidity	03/21/86	72/22.2	43	29.8/757	25.60	25.96
Max humidity Min	07/18/86	77/25.0	62	29.36/746	25.05	25.34
temperature Max	06/26/86	72/22.2	61	29.41/747	25.18	25.62
temperature Min barometric	04/04/86	80/26.7	51	29.32/745	25.14	25.17
pressure	03/19/86	73/22.8	54	28.70/729	24.63	24.95
Average weight of air displaced					25.12	25.41
Standard deviation					0.30	0.35
Maximum variation					0.95	1.02

[a] Volume of polymer tibial component: 5780-08   8 mm = 12.8 cm³; 5780-26   16 mm = 21.8 cm³.

TABLE 2—Soak control-pin weight gains (cumulative), mg.

	11/20/86	12/15/86	12/23/86	01/15/87	01/28/87	02/03/87	02/09/87	02/16/87	03/11/87
	1.79	2.17	2.21	2.39	2.71	2.87	2.94	2.76	2.95
	2.07	2.40	2.32	2.47	2.72	3.42	2.82	2.84	3.02
	1.95	2.18	2.17	2.40	2.73	2.74	2.64	3.15	2.78
	1.96	2.28	2.25	2.35	2.62	2.83	2.75	2.70	2.83
	2.14	2.32	2.18	2.40	2.78	2.85	2.64	3.70	2.77
	2.31	2.34	2.24	2.54	2.70	3.11	2.95	2.92	3.13
	1.81	1.97	1.96	2.21	2.62	2.63	2.65	2.68	2.80
	2.08	2.12	2.19	2.43	3.05	3.21	2.71	2.98	2.92
	2.12	2.54	2.38	2.58	2.94	3.29	2.95	3.11	3.03
	1.81	2.07	2.13	2.30	2.66	3.07	2.83	3.15	3.03
	2.28	2.07	2.25	2.53	2.76	3.17	3.01	3.13	2.79
	2.13	2.73	2.34	2.45	2.75	2.98	2.71	2.87	2.97
	2.00	2.24	2.18	2.45	2.90	3.20	2.95	2.86	2.97
	1.99	2.38	2.18	2.33	2.70	3.71	2.68	2.79	2.89
	1.87	2.38	2.38	2.63	2.93	3.30	3.22	3.04	2.93
	2.06	2.12	2.23	2.46	2.75	2.90	2.73	2.82	2.94
	2.11	2.19	2.31	2.52	2.72	3.11	2.85	2.98	2.83
Total days	27	52	60	83	96	102	108	115	140
Mean	2.03	2.26	2.23	2.44	2.77	3.08	2.83	2.97	2.92
Standard deviation	0.15	0.18	0.10	0.10	0.12	0.26	0.16	0.24	0.18
Barometric pressure, mm Hg	734.3	745.0	748.8	743.2	746.0	739.1	750.1	744.2	750.6
Temperature, °C	22.2	22.7	23.3	22.7	23.3	23.3	23.3	23.9	23.8
Relative humidity, %	49%	49%	46%	48%	42%	44%	42%	42%	43%
Weight of air displaced, mg	5.148	5.215	5.230	5.203	5.207	5.161	5.235	5.194	5.233
Corrected weight gain	1.948	2.245	2.23	2.41	2.75	3.01	2.83	2.93	2.92
Initial conditions, 10-24-86									
Barometric pressure, mm Hg	748.3								
Temperature, °C	23.8								
Relative humidity, %	56%								
Weight of air displaced, mg	5.230								

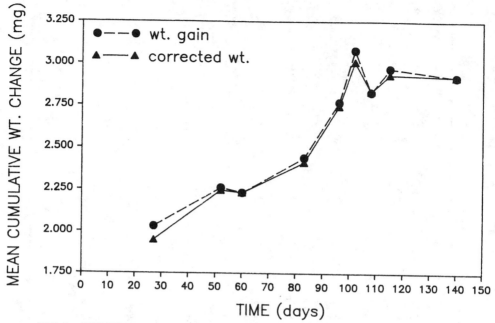

FIG. 1—*UHMWPE pin soak study graph: Mean cumulative weight change versus time.*

by the test sample. The weight of the dry air can be calculated by multiplying the displaced volume by the following equation

$$\text{Dry Air Density} = \frac{0.001293}{1 + 0.00367T} \times \frac{H}{76} \qquad \text{Ref 5}$$

Where the air density is in grams per milliliter, T is temperature in degrees centigrade and H is the barometric pressure in centimeters of mercury.

The weight of the moist air is calculated Eq from Ref 5 and is shown below. The product of the moist air density and volume of the specimen is the moist air displaced.

$$\text{Moist Air Density} = 1.2929 \cdot \frac{273.13}{T} [(H - 0.3783e)/760]$$

Where the air density is in grams per liter, T is temperature in degrees Kelvin, H is barometric pressure in millimeters of mercury, and e is the vapor pressure of the moisture in the air in millimeters of mercury. The moist air was used on all buoyancy factors. The dry air calculation is for comparison of the humidity effect.

Discussion

The purpose of a soak study of this type is to determine the rate of gain of the polymer component. The rate of gain examined was the slope of the line starting after the 30th day.

As shown on Fig. 1, the initial pin weight gains were zero. Weight gain due to fluid absorption thereafter was adjusted to the buoyancy condition shown by the solid line in Fig. 1. The buoyancy correction can be a positive or negative number depending on the initial conditions. Standard F-732 Section 6.2.5 states that polymers such as UHMW polyethylene may wear less than 100 μg per million cycles. The maximum weight variation of the pin or disk test (Table 2) due to buoyancy is 80 μg. The accuracy of weighing a specimen in an environment that changes pressure, temperature, and humidity throughout the test is buoyancy variation. A simple computer program was written to calculate the buoyancy effect on any component. The program provided quick and easy corrections to what first appeared to be many lengthy calculations.

Conclusion

Fluctuation of weight of a 35 mm by 9 mm diameter UHMWPE pin by buoyancy influence is 0.1 mg (Fig. 1). With the water absorption of the pins averaging 1000 μg/100 days, accuracy of weighing is 10% due to buoyancy variations. The slope of the weight gain curve (Fig. 1) was not changed 100 μg by the buoyancy influence. Experimental weight loss of larger specimens, such as polyethylene tibial and acetabular components, will range as high as 0.95 mg as shown in Table 1. The buoyancy force had a 0.54-mg influence on the knee test data as shown in Fig. 2. This 0.54-mg variation could equal the weight of wear debris in some tests.

Utilizing the buoyancy weight correction along with the fluid absorption correction has improved the accuracy of each individual data point. Weight loss measurements over a

KNEE WEAR TEST

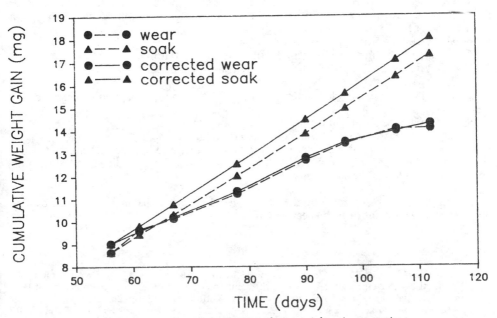

FIG 2.—*Knee wear test graph: Cumulative weight gain versus time.*

long period of time and with many data points permit averaging out the high and low buoyancy points to establish a general wear trend. Table 1 is an example of a six-month UHMWPE component which has a wear rate of 1 to 2 mg per million cycles. The weight variation is 0.3-mg standard deviation and 0.95 maximum variation. Short-term experiments with few data points could reflect atmospheric variations predominantly reflecting the trends of barometric pressure and relative humidity. Applying proper weight methodology (by subtracting out the weight gain due to the fluid absorption, using good cleaning technique, and correcting weights for the buoyancy changes), the weight loss technique is a simple and accurate method of determining wear rates. Buoyancy compensation should be considered for addition to any test procedure utilizing weight loss measurements for wear of polymeric orthopedic implants.

Acknowledgments

The authors would like to acknowledge Zimmer and its research staff for their support and guidance.

APPENDIX

Cleaning Procedure

1. All cleaning and rinsing fluids, as well as bath temperatures, will be maintained at room temperature.
2. The specimens will be cleaned in distilled water for 5 min using an ultrasonic cleaner.
3. Inspect and scrub all specimens briskly with a nylon brush to remove any dried-on bovine serum.
4. Specimens will be rinsed using a polyethylene spray bottle containing distilled water.
5. Clean specimens in a 1:4 dilution (using distilled water) detergent solution in an ultrasonic cleaner for 15 min.
6. Rinse specimen in distilled water as in Step 3.
7. Ultrasonically rinse specimen in distilled water for 10 min.
8. Using fresh distilled water, ultrasonically rinse specimen for 3 min.

NOTE: All ultrasonic distilled water rinses should be done with fresh distilled water.

9. Shake off excess water and rinse each specimen in alcohol.
10. Blow dry each specimen with compressed nitrogen.
11. Place all specimens in vacuum degassing chamber and evacuate to a pressure of 5 μm. Degas the pins for 60 min. After 60-min degas cycle, backfill the chamber with dry nitrogen and remove the specimens.

NOTE: Weigh each specimen on a Mettler AE163 analytical balance to the nearest 0.01 mg. Record the weight of each specimen.

References

[1] *1983 Annual Book of ASTM Standards,* Vol. 13.01 Medical Devices, ASTM Designation: F 732-82, Practice for Reciprocating Pin On Flat Evaluation of Friction and Wear Properties of Polymeric Materials for Use in Total Joint Prostheses, Section 13, 1983, pp. 262–269.
[2] McKellop, H., Clarke, I. C., Markolf, K., and Amstutz, H., "Wear Characteristics of UHMWPE:

A Method for Accurately Measuring Extremely Low Wear Rates," *Journal of Biomedical Materials Research,* 1978, Vol. 12, pp. 897–927.

[*3*] Clarke, I. C. and Starkebaum, W., "Fluid Absorption Phenomenon in Sterilized Polyethylene Acetabular Prostheses," *Biomaterials,* 1985, Vol. 6, pp. 184–186.

[*4*] Seeger, R. J., "Fluid Mechanics," in *Handbook of Physics,* 2nd ed., Condon and Odishaw, Eds., National Science Foundation, Washington, DC, 1967, pp. 3–15.

[*5*] *CRC Handbook of Chemistry and Physics,* 62nd ed., 1981–1982, pp. F9, F10, F11.

H. Thomas McClelland[1] and Mitchell L. Birkland[1]

A Survey of the Experimental Determination of Precision in Materials Research

REFERENCE: McClelland, H. T. and Birkland, M. L., **"A Survey of the Experimental Determination of Precision in Materials Research,"** *Factors That Affect the Precision of Mechanical Tests, ASTM STP 1025*, R. Papirno and H. C. Weiss, Eds., American Society for Testing and Materials, Philadelphia, 1989, pp. 240–242.

ABSTRACT: The purpose of this survey was to determine the extent to which the precision of data was determined during the course of experimental studies in materials research. Eight technical journals were sampled.

It was found that 23 ± 13% (at the 95% confidence level) of the papers in the journals examined reported experimentally determined precision. It would appear from these results that an effort should be made to train research personnel in modern experimental procedures.

KEY WORDS: precision, statistical research procedures, materials research

The purpose of this survey was to determine the extent to which the precision of data was determined during the course of experimental studies in materials research.

It is the opinion of the authors that the major factor affecting the precision of experimental data is the skill of the research scientist/engineer in properly planning the experimental study, and that the use of proper statistical planning and analysis methods affords the greatest probability of obtaining an accurate estimate of experimental precision. The reporting of experimental precision is essential to give a clear indication of the uncertainty involved in the data, to allow others to determine when they have reproduced the data, and to help prevent unreasonable extrapolation of the data. The study was designed to provide an overview of the current situation. It was not designed to compare individual journals so as to avoid "my journal is better than your journal" arguments.

It is anticipated that the results of this survey may form a starting point for a discussion on the need for training research personnel in modern experimental methods.

Method

The procedure followed was to examine a representative sample of technical journals in the field of materials. A statistically randomized sample of individual papers was read, and whether or not they contained information on the precision of the data was determined and recorded.

Eight technical journals were selected for the survey: (1) *Acta Metallurgica;* (2) *Scripta Metallurgica;* (3) *Journal of Material Science and Engineering;* (4) *Journal of Materials Science;* (5) *Journal of Testing and Evaluation;* (6) *Journal of the Air Pollution Control*

[1] Mechanical Engineering Department, Montana State University, Bozeman, Montana 59717–0007.

Association; (7) *Journal of Materials and Technology;* and (8) *Journal of Composite Materials.*

These journals were chosen for the following reasons: (1) they represent a cross section of materials research; (2) they are available in the Montana State University Library; and (3) they represent journals of interest to the authors. A five-year time period, January 1982 to December 1986, was chosen to reflect the current state of affairs.

A pilot study was performed to estimate the number of papers which contained experimentally determined precision values and the variation among the population. The pilot study consisted of examining three entire issues from early, middle, and late in the time period for each journal. Each paper was evaluated as to whether or not the experimental precision was given in terms of error bands (that is, xxx ± yy), error bars on graphs, or confidence limits. A statement of the precision of a particular piece of equipment was not considered sufficient. The average number of positively evaluated papers per issue and the standard deviation among the issues were calculated. The results of the pilot study were not included in the final assessment because the data selection process for the pilot study was not randomized.

Originally, it was planned to evaluate the papers as to whether or not statistical techniques were used in the planning, conduction, and analysis of the experiments, but very few experimenters provided sufficient information as to the planning or conduction stages. It was, therefore, assumed that, for the purposes of this study, any indication of experimentally determined precision would be counted as a positive value.

The results of the pilot study were used to calculate the sample size (number of papers) for the main study [*1*]. It was determined that 37 papers would be sufficient to determine the average percent of positive papers ±13% to a 95% confidence level. A total of 39 papers were actually examined, with each author evaluating one third and both authors evaluating the remaining one third of the papers.

The sample papers were selected in the following manner: the average number of papers for each issue of each journal was calculated using the results from the pilot study, and the approximate total number of papers in each journal for the time period was determined. Each of the eight journals was assigned a number from one to eight, which were then randomized. All of the papers in the total population of 7400 papers were assigned a number, and the 39 sample papers were selected using a random number generator. Table 1 contains the list of article numbers.

The numbering for each journal began with January 1982 and proceeded sequentially through December 1986. Prior to the actual data collection, it was decided that if a selected paper was strictly theoretical, the next paper would be chosen for evaluation. The articles contained in the sample population are not listed in this report but are available from the first author.

TABLE 1—*List of journals and assigned article numbers.*

Journals	Numbers
Acta Metallurgica	0 to 1079
Journal of Testing and Evaluation	1080 to 1269
Journal of Air Pollution Control Association	1270 to 1489
Scripta Metallurgica	1490 to 3379
Journal of Materials and Technology	3380 to 3639
Journal of Materials Science and Engineering	3640 to 4749
Journal of the Composite Materials	4750 to 4879
Journal of Material Science	4880 to 7399

Results

Nine of the thirty-nine articles of the sample population contained evidence of experimentally determined precision. Thus, according to our predetermined criteria, $23 \pm 13\%$ (at the 95% confidence level) of the articles in the parent population contained some indication of data scatter.

Discussion

If one assumed that the determination of the precision of experimental data is essential for technical communication, the above results are disappointing. They indicate that at least one area of the technical community is not providing sufficient information for proper communication.

The actual situation is somewhat worse than shown by the results indicated above. In all but a few cases in the pilot study and actual study, the authors who gave precision estimates did not indicate what the estimates represented, that is, whether they were a specific confidence interval, a standard error, a standard deviation, etc. Of those articles where this information was given, there was no consistency, and all of the above were used by various authors. This lack of indicating the basis for the precision measurements can lead to a serious misinterpretation of the data.

Another area of concern is the method of data collection. In only two articles of the over three hundred examined did the authors clearly indicate that the data collection was conducted in a statistically randomized manner. Failure to follow this procedure can lead to incorrect measures of precision, usually showing better precision than is actually present.

The overall results of this study indicate that there is a serious lack of knowledge of modern experimental techniques among a large portion of the research personnel currently publishing. Since many of these personnel are probably university faculty members, it is probable that our future researchers are also not learning these techniques.

Conclusions

The percentage of articles in the parent population of eight journals in the time period of January 1982 through December 1986 containing evidence of experimentally determined precision of the data was $23 \pm 13\%$ at the 95% confidence level. It would appear from these results that an effort should be made to train research personnel in modern experimental techniques.

Reference

[1] Mandel, J., *The Statistical Analysis of Experimental Data,* Dover Publishing, p. 233, 1984 (Copyright 1964).

Author Index

Subject Index